W9-AZH-466

Learn ◆ Practice ◆ Succeed

Eureka Math® student materials for *A Story of Units*® (K–5) are available in the *Learn, Practice, Succeed* trio. This series supports differentiation and remediation while keeping student materials organized and accessible. Educators will find that the *Learn, Practice,* and *Succeed* series also offers coherent—and therefore, more effective—resources for Response to Intervention (RTI), extra practice, and summer learning.

Learn

Eureka Math Learn serves as a student's in-class companion where they show their thinking, share what they know, and watch their knowledge build every day. *Learn* assembles the daily classwork—Application Problems, Exit Tickets, Problem Sets, templates—in an easily stored and navigated volume.

Practice

Each *Eureka Math* lesson begins with a series of energetic, joyous fluency activities, including those found in *Eureka Math Practice.* Students who are fluent in their math facts can master more material more deeply. With *Practice,* students build competence in newly acquired skills and reinforce previous learning in preparation for the next lesson.

Together, *Learn* and *Practice* provide all the print materials students will use for their core math instruction.

Succeed

Eureka Math Succeed enables students to work individually toward mastery. These additional problem sets align lesson by lesson with classroom instruction, making them ideal for use as homework or extra practice. Each problem set is accompanied by a Homework Helper, a set of worked examples that illustrate how to solve similar problems.

Teachers and tutors can use *Succeed* books from prior grade levels as curriculum-consistent tools for filling gaps in foundational knowledge. Students will thrive and progress more quickly as familiar models facilitate connections to their current grade-level content.

Students, families, and educators:

Thank you for being part of the *Eureka Math*® community, where we celebrate the joy, wonder, and thrill of mathematics.

Nothing beats the satisfaction of success—the more competent students become, the greater their motivation and engagement. The *Eureka Math Succeed* book provides the guidance and extra practice students need to shore up foundational knowledge and build mastery with new material.

What is in the Succeed *book?*

Eureka Math Succeed books deliver supported practice sets that parallel the lessons of *A Story of Units*®. Each *Succeed* lesson begins with a set of worked examples, called *Homework Helpers*, that illustrate the modeling and reasoning the curriculum uses to build understanding. Next, students receive scaffolded practice through a series of problems carefully sequenced to begin from a place of confidence and add incremental complexity.

How should Succeed *be used?*

The collection of *Succeed* books can be used as differentiated instruction, practice, homework, or intervention. When coupled with *Affirm*®, *Eureka Math*'s digital assessment system, *Succeed* lessons enable educators to give targeted practice and to assess student progress. *Succeed*'s perfect alignment with the mathematical models and language used across *A Story of Units* ensures that students feel the connections and relevance to their daily instruction, whether they are working on foundational skills or getting extra practice on the current topic.

Where can I learn more about Eureka Math *resources?*

The Great Minds® team is committed to supporting students, families, and educators with an ever-growing library of resources, available at eureka-math.org. The website also offers inspiring stories of success in the *Eureka Math* community. Share your insights and accomplishments with fellow users by becoming a *Eureka Math* Champion.

Best wishes for a year filled with Eureka moments!

Jill Diniz

Jill Diniz
Director of Mathematics
Great Minds

Contents

Module 4: Number Pairs, Addition and Subtraction to 10

Module 5: Numbers 10–20 and Counting to 100

© 2018 Great Minds® eureka-math.org

Module 6: Analyzing, Comparing, and Composing Shapes

Grade K
Module 4

Number Bonds

Number bonds are models that show how numbers can be taken apart. The bigger number is the *whole*, or *total*, and the smaller numbers are the *parts* except when there is a 0. For now, please use everyday words such as "is," "and," and "make." Addition and subtraction will come later in this module. Number bonds are shown in different positions so that students can become flexible thinkers!

Draw the dark butterflies in the first circle on top. Draw the light butterflies in the next circle on top. Draw all the butterflies in the bottom circle.

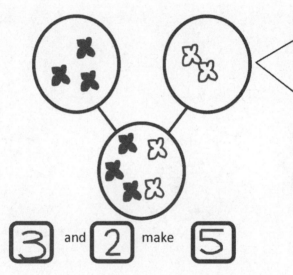

There are 3 dark butterflies; that's one part.

There are 2 light butterflies; that's the other part.

When I count them all, there are 5. That's the total, or whole.

Lesson 1:   Model composition and decomposition of numbers to 5 using actions, objects, and drawings.

© 2018 Great Minds®. eureka-math.org

3

Name _____   Date _____

Draw the blue fish in the first circle on top. Draw the orange fish in the next circle on top. Draw all the fish in the bottom circle.

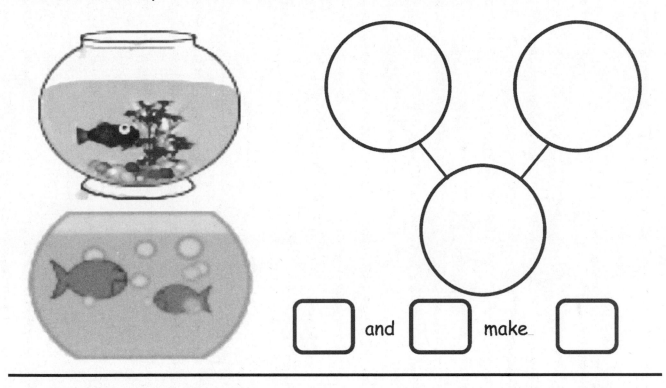

☐ and ☐ make ☐

Draw a square for each fish in the top circle. Draw a square for each goldfish in the bottom circle. In the last circle on the bottom, draw a square for each spiny fish.

☐ and ☐ make ☐

EUREKA MATH

Lesson 1:   Model composition and decomposition of numbers to 5 using actions, objects, and drawings.

© 2018 Great Minds®. eureka-math.org

5

The squares below represent a cube stick. Color some squares blue and the rest of the squares red. Draw the squares you colored in the number bond. Show the hidden partners on your fingers to an adult. Color the fingers you showed.

I decided to color 4 squares blue and 1 red.
I could have colored 3 and 2. Any way I color, there are 5 squares in all.

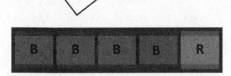

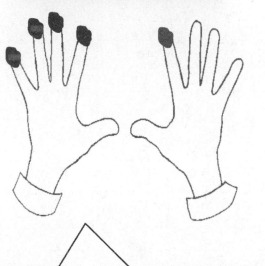

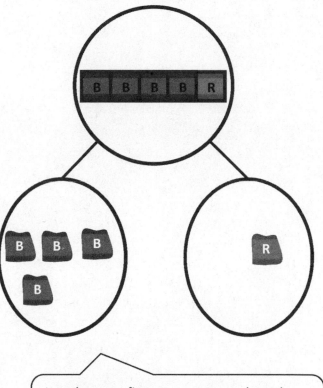

I show 4 fingers on one hand and 1 on the other hand. That's 5 fingers in all.

Here are the fingers I showed. Can you think of another way?

I see how my fingers, squares, and number bond match: 4 and 1 make 5. I can also say 5 is the same as 4 and 1.

EUREKA
MATH®

Lesson 2: Model composition and decomposition of numbers to 5 using fingers and linking cube sticks.

© 2018 Great Minds®. eureka-math.org

7

Name _____ Date _____

The squares below represent a cube stick. Color some squares blue and the rest of the squares red. Draw the squares you colored in the number bond. Show the hidden partners on your fingers to an adult. Color the fingers you showed.

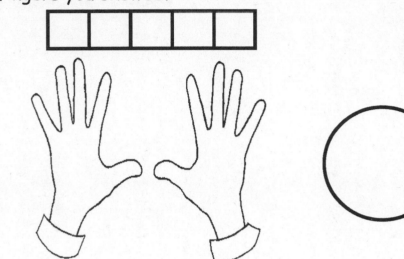

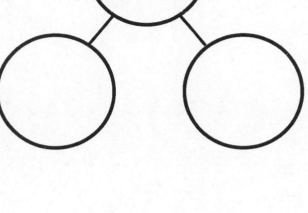

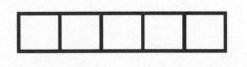

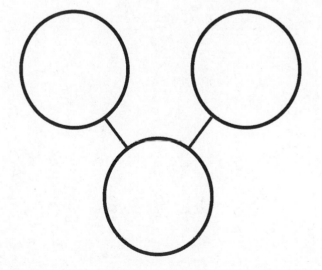

Lesson 2: Model composition and decomposition of numbers to 5 using fingers and linking cube sticks.

© 2018 Great Minds®. eureka-math.org

9

Fill in the number bond to match the domino.

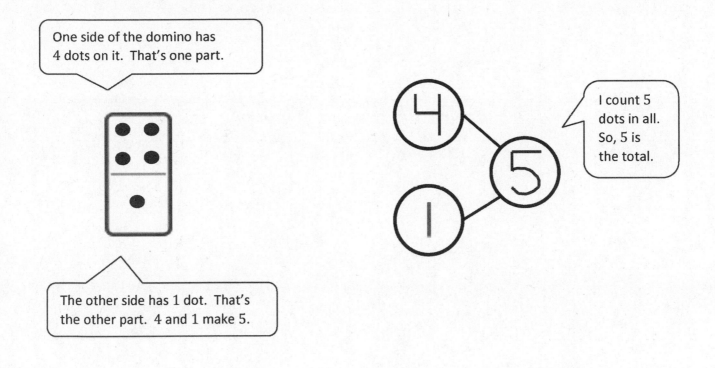

One side of the domino has 4 dots on it. That's one part.

I count 5 dots in all. So, 5 is the total.

The other side has 1 dot. That's the other part. 4 and 1 make 5.

Fill in the domino with dots, and fill in the number bond to match.

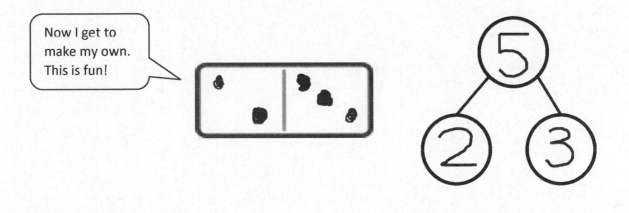

Now I get to make my own. This is fun!

EUREKA MATH

Lesson 3:   Represent composition story situations with drawings using numeric number bonds.

© 2018 Great Minds®. eureka-math.org

11

Name _____   Date _____

Fill in the number bond to match the domino.

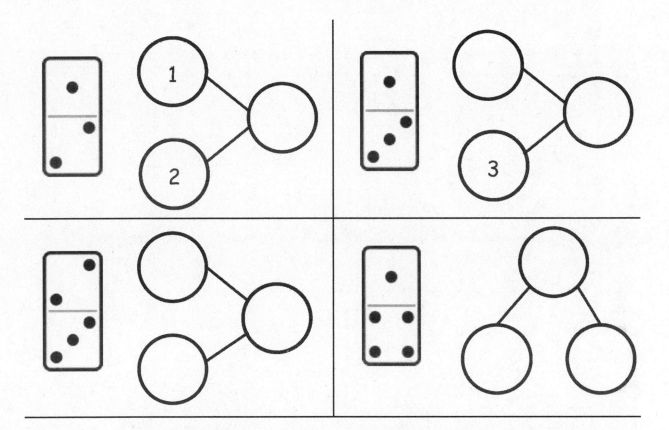

Fill in the domino with dots, and fill in the number bond to match.

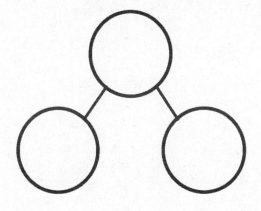

Lesson 3:   Represent composition story situations with drawings using numeric
number bonds.

© 2018 Great Minds®. eureka-math.org

13

Finish the number bond. Finish the sentence.

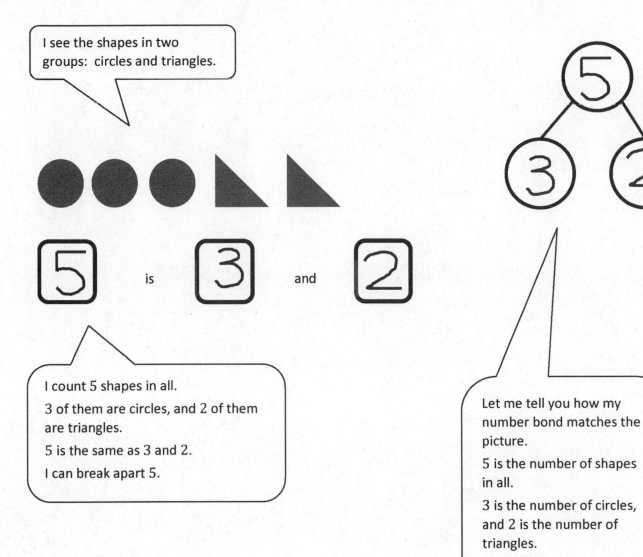

I see the shapes in two groups: circles and triangles.

5 is 3 and 2

I count 5 shapes in all.

3 of them are circles, and 2 of them are triangles.

5 is the same as 3 and 2.

I can break apart 5.

Let me tell you how my number bond matches the picture.

5 is the number of shapes in all.

3 is the number of circles, and 2 is the number of triangles.

I can break apart 5.

 EUREKA MATH

Lesson 4: Represent decomposition story situations with drawings using numeric number bonds.

© 2018 Great Minds®. eureka-math.org

15

Name _____   Date _____

Finish the number bonds. Finish the sentence.

3 is ☐ and ☐

2   1

☐ is ☐ and ☐

4

Tell an adult a story about the animals, and then make a number sentence and number bond about it.

☐ is ☐ and ☐

Lesson 4: Represent decomposition story situations with drawings using numeric number bonds.

17

© 2018 Great Minds®. eureka-math.org

EUREKA MATH

Tell a story about the picture. Fill in the number bond and the sentence to match your story.

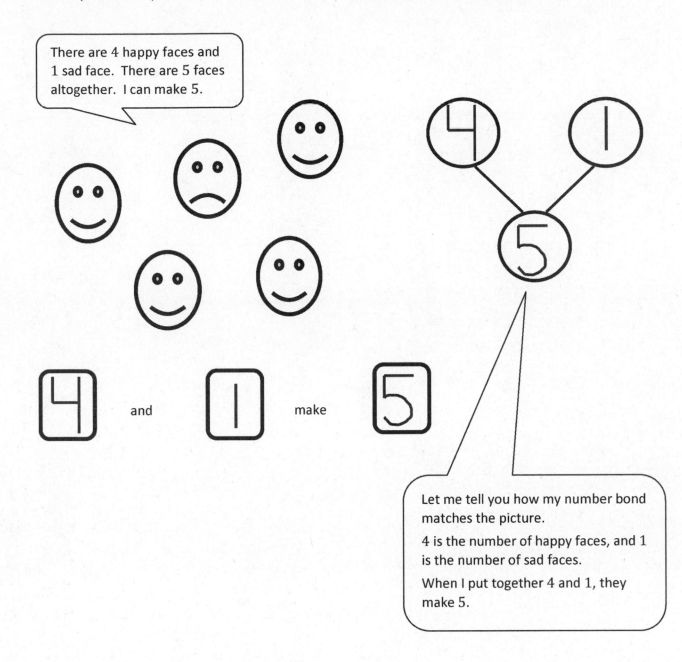

and   make

Lesson 5:   Represent composition and decomposition of numbers to 5 using
pictorial and numeric number bonds.

© 2018 Great Minds®. eureka-math.org

19

Name _____ Date _____

There are 2 pandas in a tree. 2 more are walking on the ground. How many pandas are there? Fill in the number bond and the sentence.

[ ] and [ ] make [ ]

Tell a story about the penguins. Fill in the number bond and the sentence to match your story.

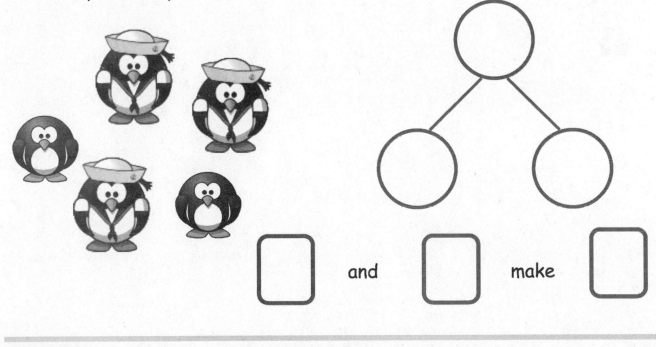

[ ] and [ ] make [ ]

Lesson 5:   Represent composition and decomposition of numbers to 5 using pictorial and numeric number bonds.

21

© 2018 Great Minds®. eureka-math.org

Tell a story. Complete the number bond. Draw pictures that match your story and number bond.

Draw some animals for your story.

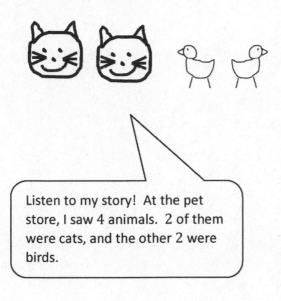

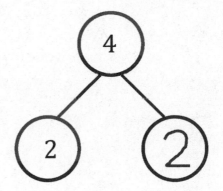

Listen to my story! At the pet store, I saw 4 animals. 2 of them were cats, and the other 2 were birds.

4 is the total. So I draw 4 animals.

2 is one of the parts, so I draw 2 cats.

The other part must be 2 since 4 is the same as 2 and 2.

I draw 2 birds to make 4 animals in all.

Name _____   Date _____

Tell a story. Complete the number bonds. Draw pictures that match your story and number bonds

Draw some balls for your story.

3

2

Draw some crayons for your story.

4

Draw some shapes for your story.

On the back of your paper, draw a picture and make a number bond.

Lesson 6:   Represent number bonds with composition and decomposition story situations.

25

© 2018 Great Minds®. eureka-math.org

Look at the shapes. Make 2 different number bonds. Tell an adult about the numbers you put in the number bonds.

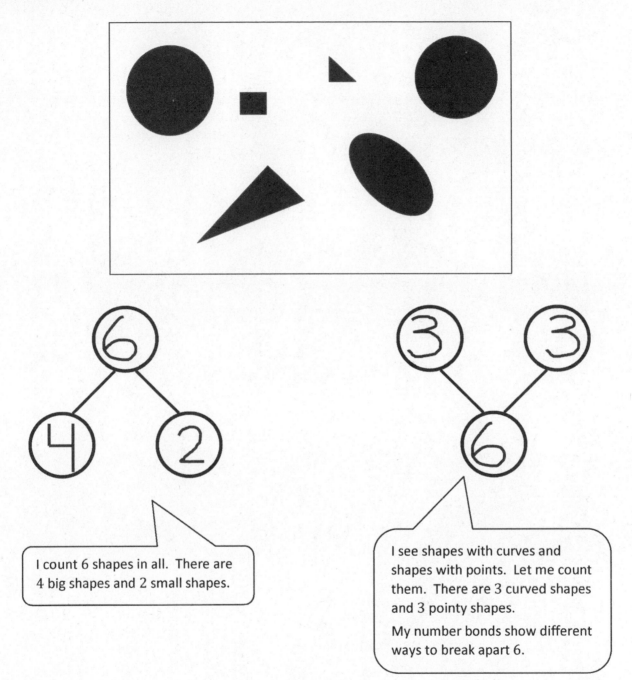

I count 6 shapes in all. There are 4 big shapes and 2 small shapes.

I see shapes with curves and shapes with points. Let me count them. There are 3 curved shapes and 3 pointy shapes.

My number bonds show different ways to break apart 6.

Name _____   Date _____

Look at the presents. Make 2 different number bonds. Tell an adult about the numbers you put in the number bonds.

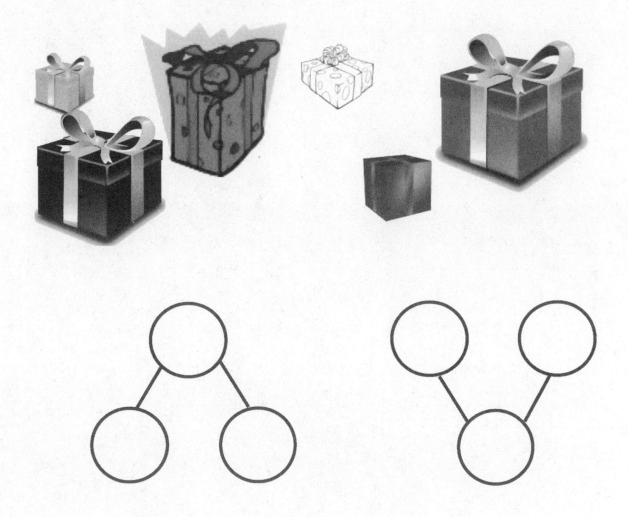

On the back of your paper, draw 6 presents, and sort them into 2 groups. Make a number bond, and fill it in according to your sort.

Lesson 7:   Model decompositions of 6 using a story situation, objects, and number bonds.

© 2018 Great Minds®. eureka-math.org

29

The squares represent cube sticks. Color some cubes red and the rest blue. Fill in the number bond and sentence to match.

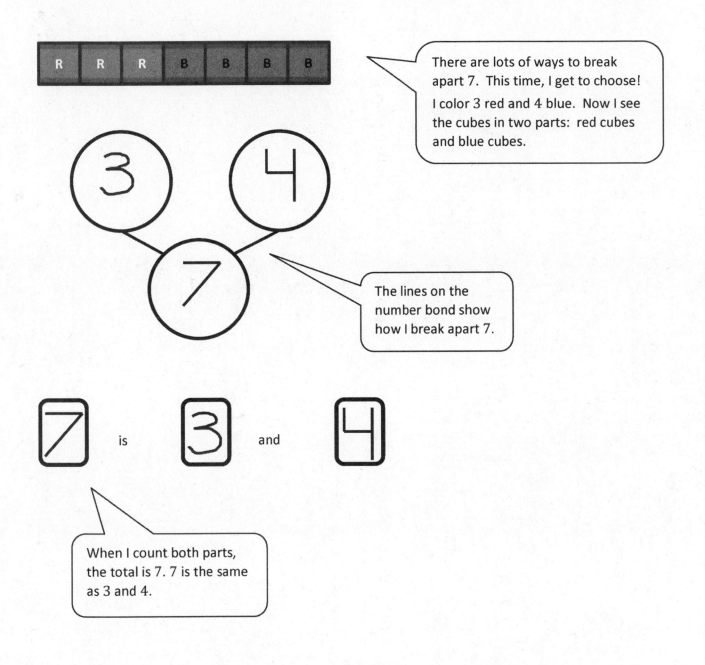

There are lots of ways to break apart 7. This time, I get to choose!

I color 3 red and 4 blue. Now I see the cubes in two parts: red cubes and blue cubes.

The lines on the number bond show how I break apart 7.

When I count both parts, the total is 7. 7 is the same as 3 and 4.

Lesson 8:   Model decompositions of 7 using a story situation, sets, and number bonds.

© 2018 Great Minds®. eureka-math.org

31

Name _____   Date _____

Draw a set of 4 circles and 3 triangles. How many shapes do you have?
Fill in the number sentence and number bond.

☐   is   ☐   and   ☐

The squares represent cube sticks. Color
the cubes to match the number bond.

7   5   2

Color some cubes red and the rest blue.
Fill out the number bond to match.

On the back of your paper, draw a set of 7 squares and circles. Make a number bond,
and fill it in. Now, write a number sentence like the sentence above that
tells about your set.

EUREKA
MATH®

Lesson 8:   Model decompositions of 7 using a story situation, sets, and
             number bonds.

© 2018 Great Minds®. eureka-math.org

33

Complete the number bond to match the domino.

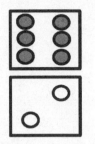   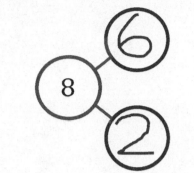

Let me tell you how my number bond and domino match.

8 tells how many dots in all.

6 is the number of grey dots.

2 is the number of white dots.

Draw a line to make 2 groups of dots. Fill in the number bond.

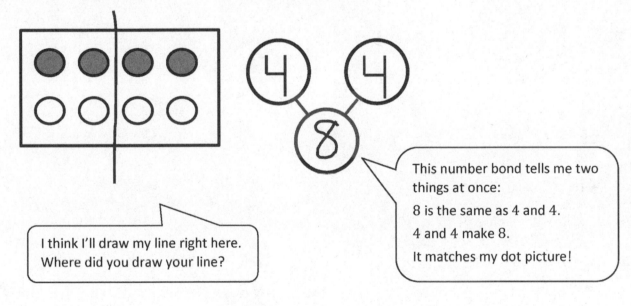

I think I'll draw my line right here. Where did you draw your line?

This number bond tells me two things at once:

8 is the same as 4 and 4.

4 and 4 make 8.

It matches my dot picture!

Lesson 9:   Model decompositions of 8 using a story situation, arrays, and
                 number bonds.

© 2018 Great Minds®. eureka-math.org

35

The squares below represent cubes. Color 7 cubes green and 1 blue. Fill in the number bond.

8 is 7 and 1

The whole stick has 8 cubes. The parts are 7 and 1.

Color 6 cubes green and 2 blue. Fill in the number bond.

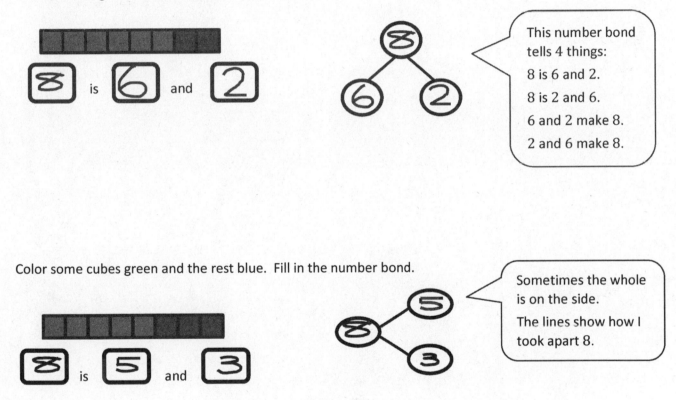

8 is 6 and 2

This number bond tells 4 things:

8 is 6 and 2.

8 is 2 and 6.

6 and 2 make 8.

2 and 6 make 8.

Color some cubes green and the rest blue. Fill in the number bond.

8 is 5 and 3

Sometimes the whole is on the side.

The lines show how I took apart 8.

Name _____   Date _____

These squares below represent cubes. Color 7 cubes green and 1 blue. Fill in the number bond.

[ ][ ][ ][ ][ ][ ][ ][ ]

[ ] is [ ] and [ ]

Color 6 cubes green and 2 blue. Fill in the number bond.

[ ][ ][ ][ ][ ][ ][ ][ ]

[ ] is [ ] and [ ]

Color some cubes green and the rest blue. Fill in the number bond.

[ ][ ][ ][ ][ ][ ][ ][ ]

[ ] is [ ] and [ ]

Color 4 cubes green and 4 blue. Fill in the number bond.

☐☐☐☐☐☐☐☐

⬜ is ⬜ and ⬜

Color 3 cubes green and 5 blue. Fill in the number bond.

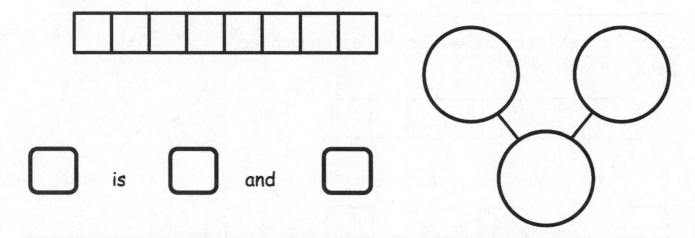

⬜ is ⬜ and ⬜

Color some cubes green and the rest blue. Fill in the number bond.

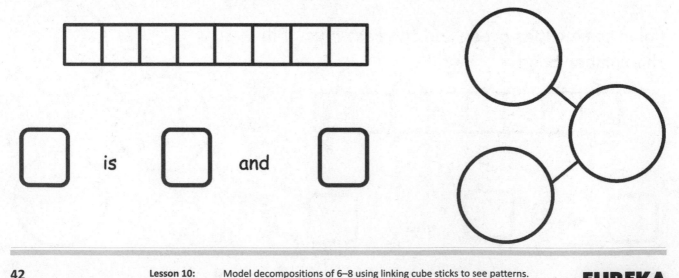

⬜ is ⬜ and ⬜

42   Lesson 10:   Model decompositions of 6–8 using linking cube sticks to see patterns.

© 2018 Great Minds®. eureka-math.org

EUREKA MATH

These squares represent cubes. Color 5 cubes green and 1 blue. Fill in the number bond.

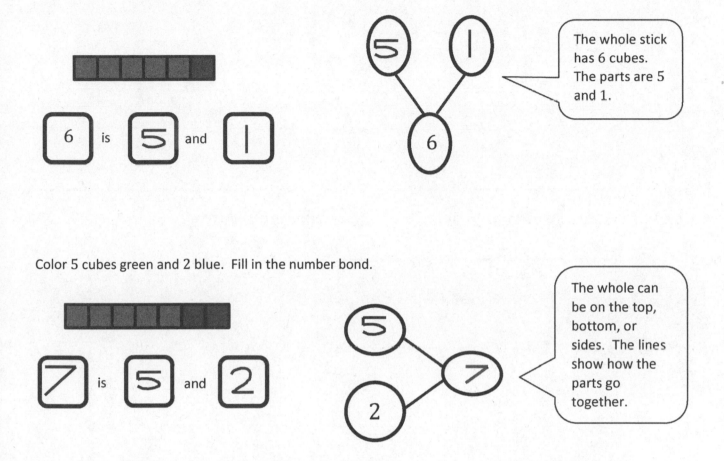

The whole stick has 6 cubes. The parts are 5 and 1.

Color 5 cubes green and 2 blue. Fill in the number bond.

The whole can be on the top, bottom, or sides. The lines show how the parts go together.

Name _____   Date _____

These squares represent cubes. Color 5 cubes green and
1 blue. Fill in the number bond.

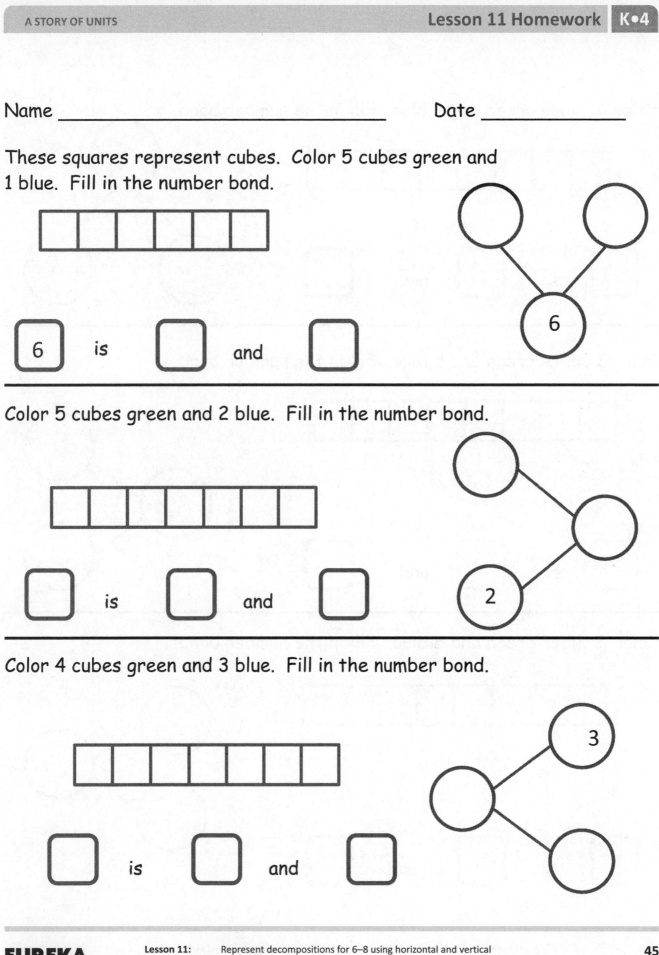

6 is ☐ and ☐

6

Color 5 cubes green and 2 blue. Fill in the number bond.

☐ is ☐ and ☐

2

Color 4 cubes green and 3 blue. Fill in the number bond.

☐ is ☐ and ☐

3

EUREKA MATH

Lesson 11:   Represent decompositions for 6–8 using horizontal and vertical
number bonds.

© 2018 Great Minds®. eureka-math.org

45

Color 4 cubes green and 4 blue. Fill in the number bond.

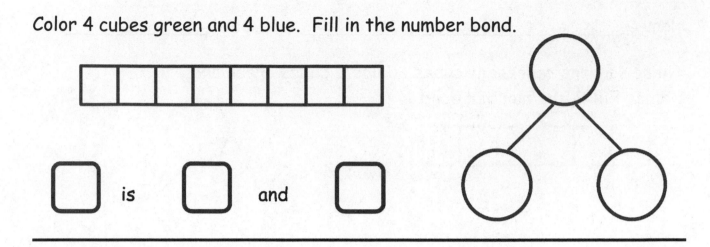

☐ is ☐ and ☐

Color 3 cubes green and 5 blue. Fill in the number bond.

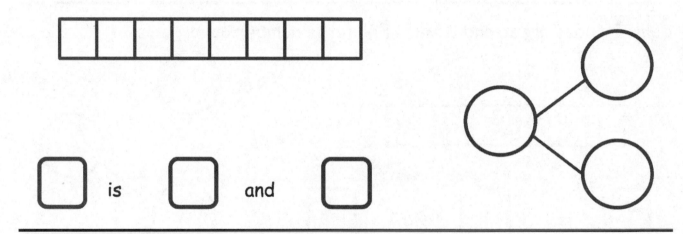

☐ is ☐ and ☐

Color 2 cubes green and 6 blue. Fill in the number bond.

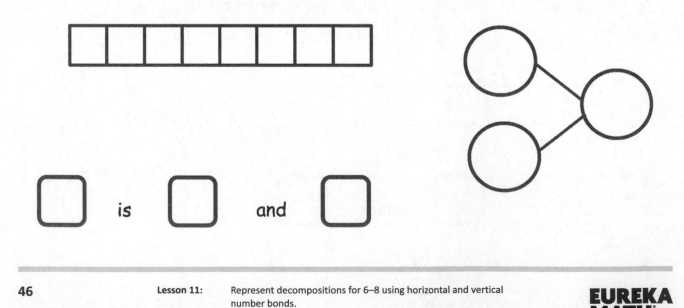

☐ is ☐ and ☐

Lesson 11:   Represent decompositions for 6–8 using horizontal and vertical
                        number bonds.

© 2018 Great Minds®. eureka-math.org

EUREKA
MATH

Fill in the number bond to match the squares.

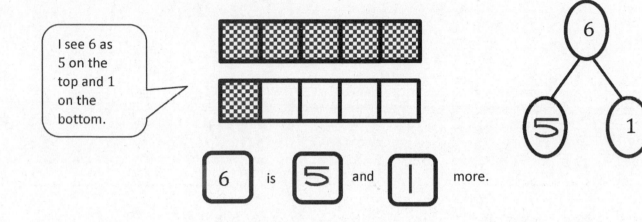

I see 6 as 5 on the top and 1 on the bottom.

6 is 5 and 1 more.

Color 5 squares blue in the first row.

Color 2 squares red in the second row.

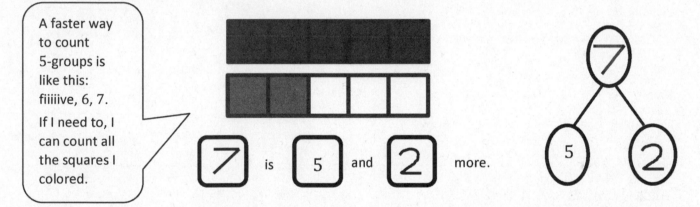

A faster way to count 5-groups is like this: fiiiiive, 6, 7.

If I need to, I can count all the squares I colored.

7 is 5 and 2 more.

Name _____   Date _____

Fill in the number bond to match the squares.

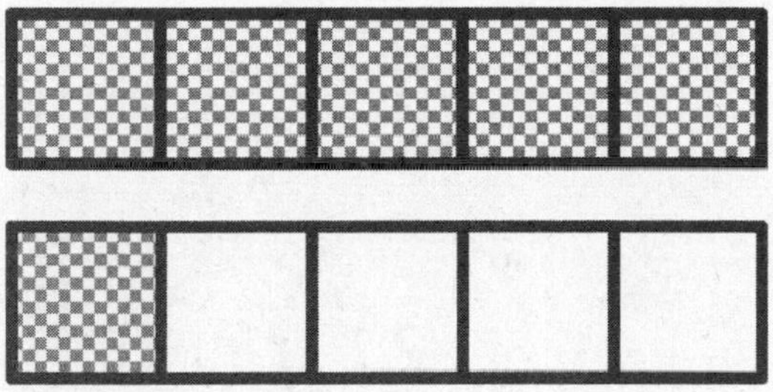

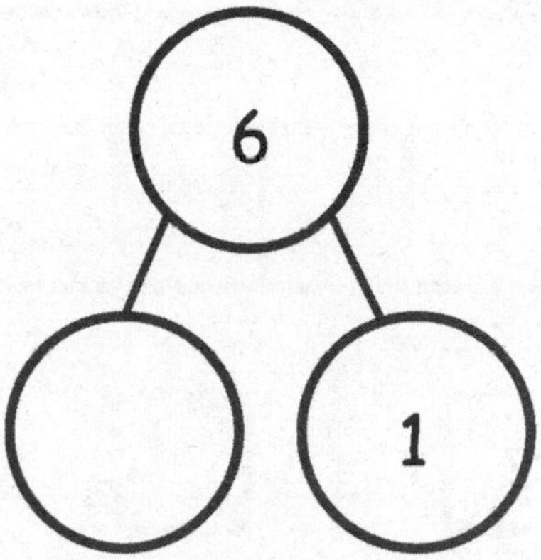

6 is ☐ and **1** more

(Number bond: 6 at top, branches to empty circle and 1)

Color 5 squares blue in the first row.

Color 2 squares red in the second row.

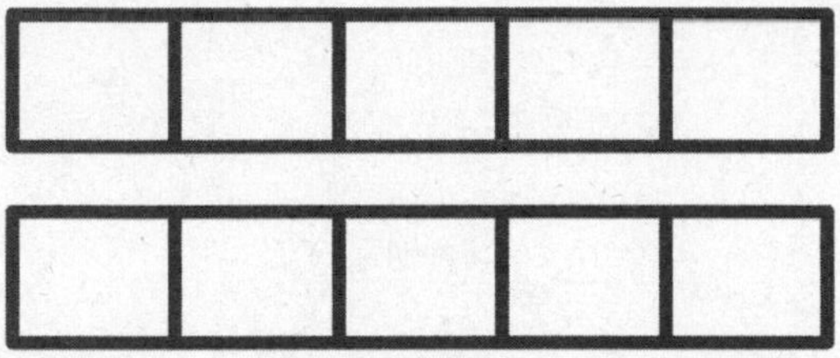

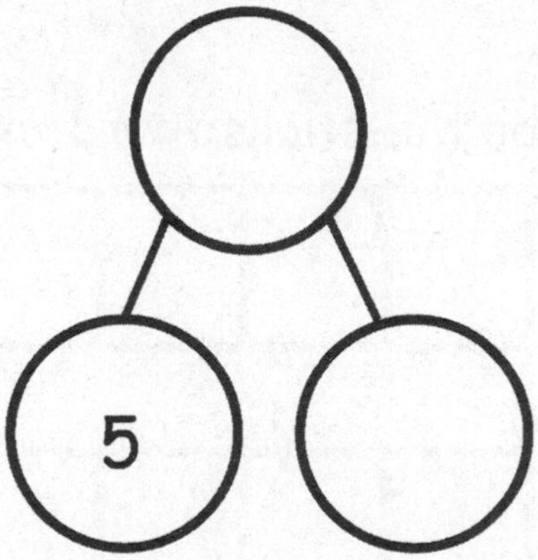

☐ is **5** and ☐ more

(Number bond: empty top circle, branches to 5 and empty circle)

Color 8 squares. Complete the number bond and sentence.

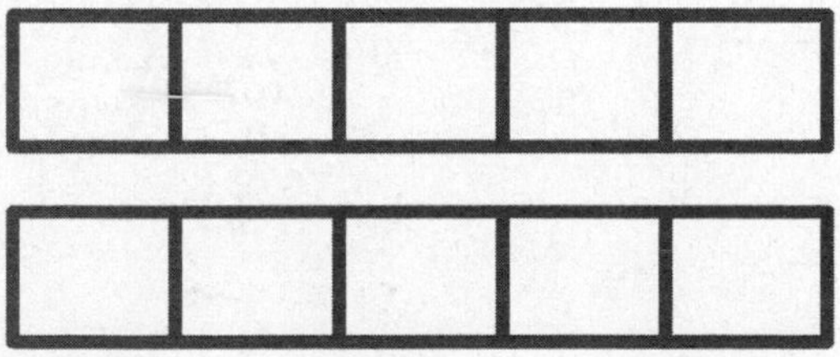

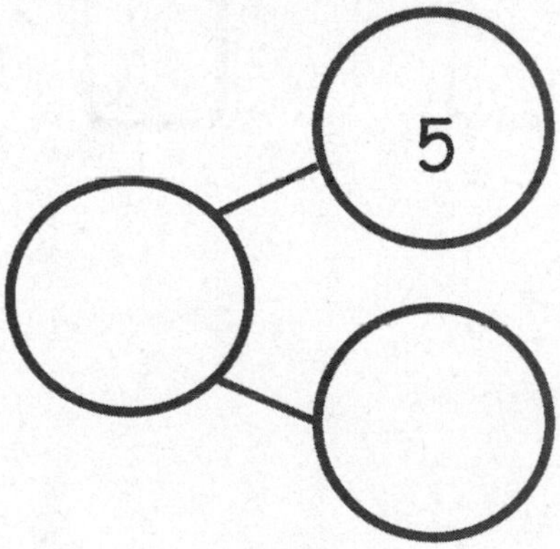

☐ is **5** and ☐ more

(Number bond: 5 at top right, branches to two empty circles)

Color the sticks to match the number bond.

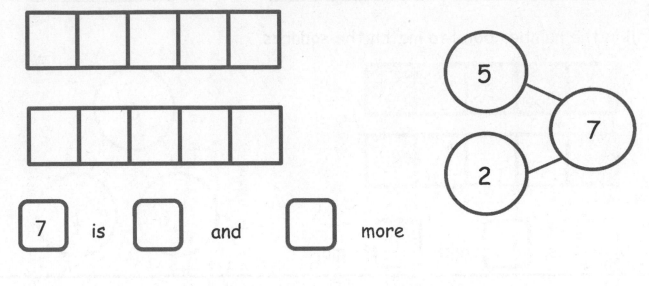

7 is ⬜ and ⬜ more

Color the sticks to match the number bond.

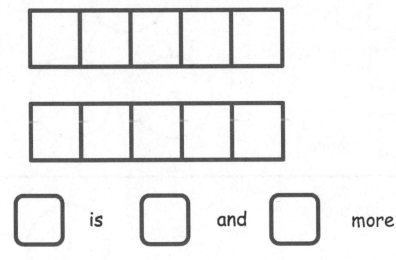

⬜ is ⬜ and ⬜ more

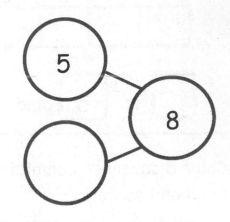

Lesson 12:   Use 5-groups to represent the 5 + *n* pattern to 8.

© 2018 Great Minds®. eureka-math.org

There are 3 monkeys and 3 elephants. All 6 animals are going into the circus tent. Fill in the number sentence and the number bond.

This story starts with the parts and ends with the whole.

I'll write my number sentences that way, too!

| 3 | and | 3 | is | 6 |

| 3 | + | 3 | = | 6 |

There are 6 animals. 4 are tigers, and 2 are lions. Fill in the number sentences and the number bond.

This story is different. It starts with the whole and ends with the parts.

I'll write my number sentences that way, too!

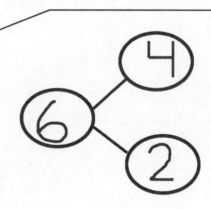

| 6 | is | 4 | and | 2 |

| 6 | = | 4 | + | 2 |

EUREKA MATH

Lesson 13:   Represent decomposition and composition addition stories to 6 with drawings and equations with no unknown.

© 2018 Great Minds®. eureka-math.org

51

Name _____   Date _____

There are 6 animals. 4 are tigers, and 2 are lions.
Fill in the number sentences and the number bond.

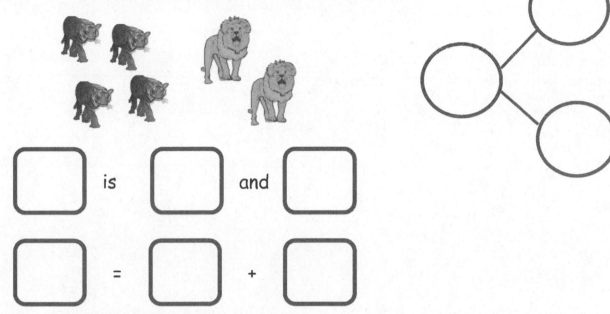

There 3 are monkeys and 3 elephants. All 6 animals are going into the
circus tent. Fill in the number sentences and the number bond.

On the back of your paper, draw some animals. Make a number bond
to match your picture.

Lesson 13:   Represent decomposition and composition addition stories to 6 with
               drawings and equations with no unknown.

© 2018 Great Minds®. eureka-math.org

53

There are 7 bears. 3 bears have bowties. 4 bears have hearts. Fill in the number sentences and the number bond.

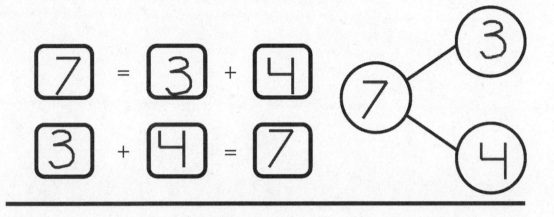

I wrote the addition sentences both ways: take apart and put together. My number bond shows that, too!

Lesson 14:   Represent decomposition and composition addition stories to 7
                   with drawings and equations with no unknown.   55

© 2018 Great Minds®. eureka-math.org

Name _____   Date _____

There are 7 bears. 3 bears have bowties. 4 bears have hearts. Fill in the number sentences and the number bond.

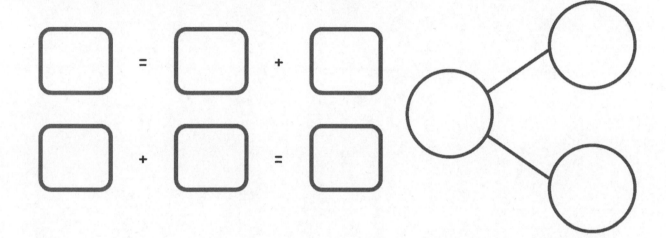

5 bears have scarves on, and 2 do not. There are 7 bears.
Write a number sentence that tells about the bears.

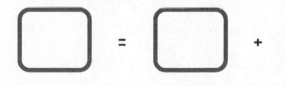

On the back of your paper, draw a picture about the 7 bears. Write a number sentence, and make a number bond to go with it.

Lesson 14:   Represent decomposition and composition addition stories to 7
with drawings and equations with no unknown.

57

© 2018 Great Minds®. eureka-math.org

There are 8 trees. 5 are palm trees, and 3 are apple trees. Fill in the number sentences and the number bond.

This addition sentence shows that there are 8 trees: 5 of one kind and 3 of another.

This addition sentence shows how the parts go together to make 8.

8 is the whole.
5 and 3 are the parts.

Lesson 15:   Represent decomposition and composition addition stories to 8 with drawings and equations with no unknown.

© 2018 Great Minds®. eureka-math.org

59

Name _____   Date _____

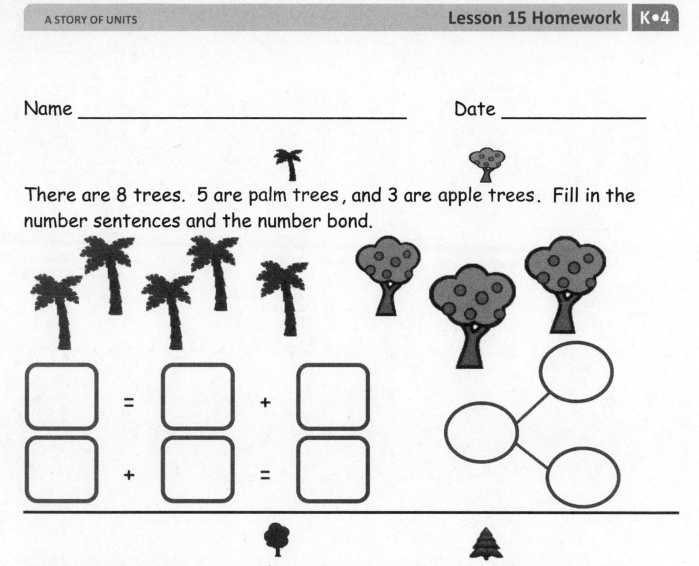

There are 8 trees. 5 are palm trees, and 3 are apple trees. Fill in the number sentences and the number bond.

There are 8 trees. 4 are oak trees, and 4 are spruce trees. Fill in the number sentences and the number bond.

Lesson 15:   Represent decomposition and composition addition stories to 8 with drawings and equations with no unknown.

61

© 2018 Great Minds®. eureka-math.org

There are 3 penguins on the ice. 4 more penguins are coming. How many penguins are there?

To find the mystery number, I can count all of the penguins: 1, 2, 3, 4, 5, 6, 7. There are 7 penguins in all!

The mystery box is for the number we don't know. I can trace the mystery box.

$$3 + 4 = \boxed{7}$$

Lesson 16:   Solve *add to with result unknown* word problems to 8 with equations.
             Box the unknown.

© 2018 Great Minds®. eureka-math.org

63

Name _____   Date _____

There are 3 penguins on the ice.
4 more penguins are coming.
How many penguins are there?

3   +   4   =   [ ]

There is 1 mama bear. 5 baby
bears are following her. How many
bears are there? Draw a box for
the answer.

1   +   5   =

Draw 7 balls in the ball box. Draw a girl putting 1 more ball in the
ball box. Circle all the balls, and draw a box for the answer. Write
your answer.

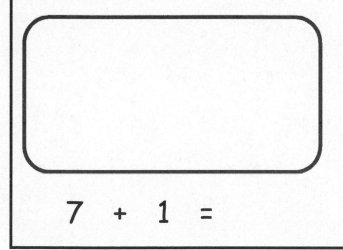

7   +   1   =

Lesson 16:   Solve *add to with result unknown* word problems to 8 with equations.
Box the unknown.

© 2018 Great Minds®. eureka-math.org

65

There are 5 hexagons and 2 triangles. How many shapes are there?

I can add the hexagons and the triangles.

The total number of shapes is 7.

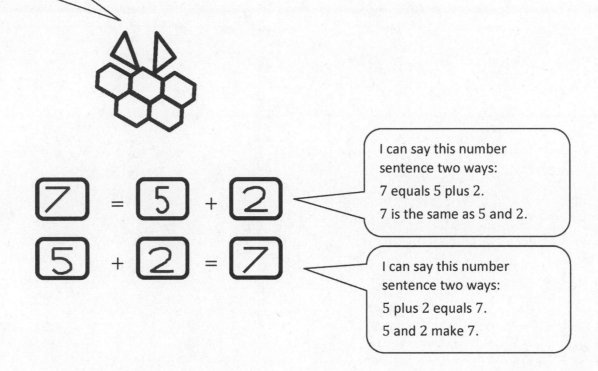

I can say this number sentence two ways:

7 equals 5 plus 2.

7 is the same as 5 and 2.

I can say this number sentence two ways:

5 plus 2 equals 7.

5 and 2 make 7.

Name _____   Date _____

There are 5 hexagons and 2 triangles. How many shapes are there?

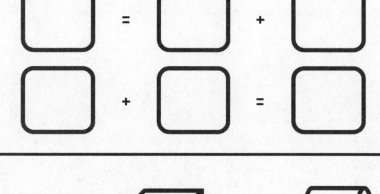

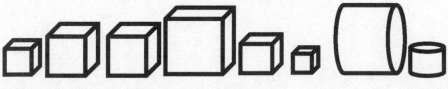

There are 6 cubes and 2 cylinders. How many shapes are there?

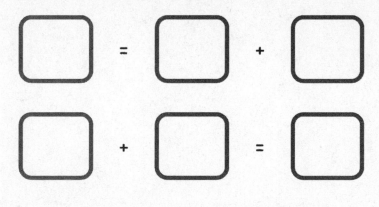

On the back of your paper, draw some shapes, and make a number sentence to match.

Lesson 17:   Solve *put together with total unknown* word problems to 8 using objects and drawings.

69

© 2018 Great Minds®. eureka-math.org

Devin has 6 pencils. He put some in his desk and the rest in his pencil box. Write a number sentence to show how many pencils Devin might have in his desk and pencil box.

> The total is 6.
> I get to choose how many of each!

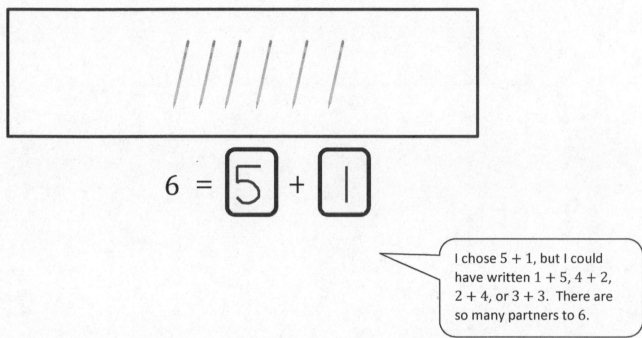

$$6 = \boxed{5} + \boxed{1}$$

> I chose $5 + 1$, but I could have written $1 + 5$, $4 + 2$, $2 + 4$, or $3 + 3$. There are so many partners to 6.

EUREKA MATH

Lesson 18: Solve *both addends unknown* word problems to 8 to find addition patterns in number pairs.

71

© 2018 Great Minds®. eureka-math.org

Name _____   Date _____

Ted has 7 toy cars. Color some cars red and the rest blue. Write a number sentence that shows how many are red and how many are blue.

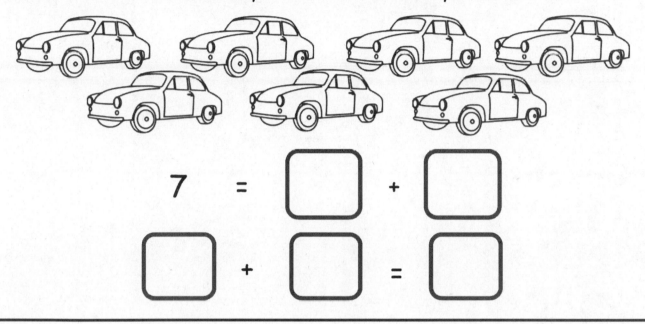

7 = [   ] + [   ]

[   ] + [   ] = [   ]

Chuck has 8 balls. Some are red, and the rest are blue. Color to show Chuck's balls. Fill in the number sentences.

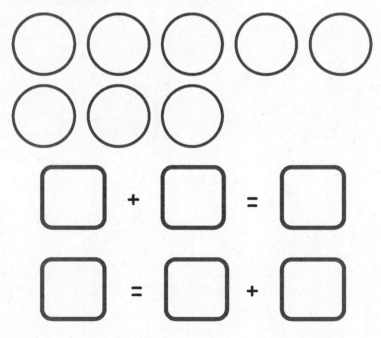

[   ] + [   ] = [   ]

[   ] = [   ] + [   ]

EUREKA
MATH

Lesson 18: Solve *both addends unknown* word problems to 8 to find addition patterns in number pairs.

73

© 2018 Great Minds®. eureka-math.org

Later I'll learn about "minus." For now, I can say that 5 trains take away 1 train is 4 trains.

1 train drove away. Cross out 1. Write how many were left.

4 tells how many are left.

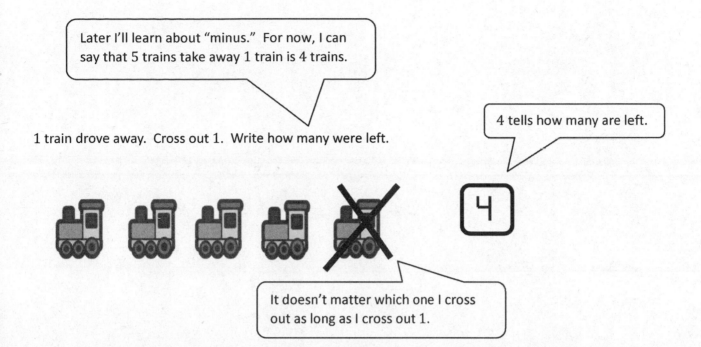

4

It doesn't matter which one I cross out as long as I cross out 1.

Two Ways to Cross Out

One at a time

All at once

EUREKA
MATH®

© 2018 Great Minds®. eureka-math.org

Name _____   Date _____

1 train drove away. Cross out 1. Write how many were left.

2 horses were bought. Cross out 2. How many were left at the store?

4 ducks swam away. Cross out 4. Write how many are left.

There are 7 apples in the tree. Draw them. A bird ate 1 of them, so cross it out. How many apples are left?

The squares below represent cube sticks. Match the cube stick to the number sentence.

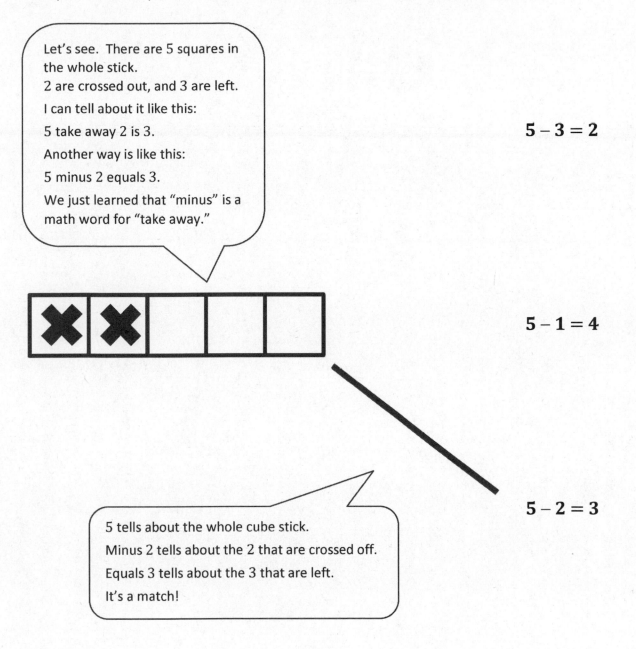

Let's see. There are 5 squares in the whole stick.

2 are crossed out, and 3 are left.

I can tell about it like this:

5 take away 2 is 3.

Another way is like this:

5 minus 2 equals 3.

We just learned that "minus" is a math word for "take away."

$$5 - 3 = 2$$

$$5 - 1 = 4$$

$$5 - 2 = 3$$

5 tells about the whole cube stick.

Minus 2 tells about the 2 that are crossed off.

Equals 3 tells about the 3 that are left.

It's a match!

Name _____   Date _____

The squares below represent cube sticks. Match the cube stick to the number sentence.

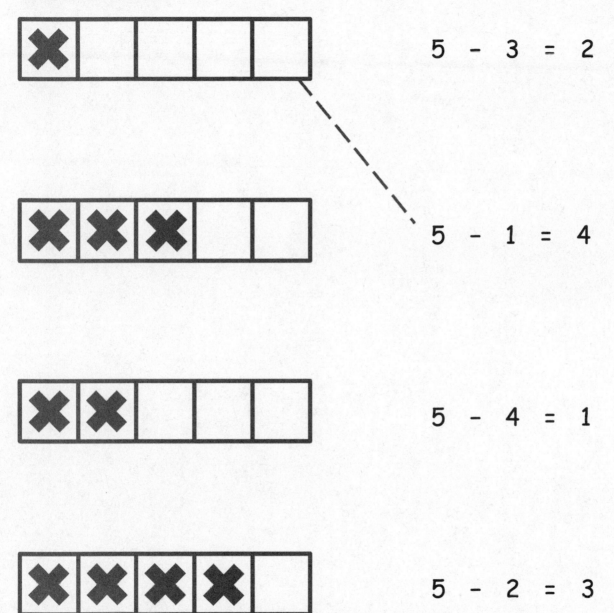

5 – 3 = 2

5 – 1 = 4

5 – 4 = 1

5 – 2 = 3

On the back of the paper, draw a 5-stick, cross out some cubes, and write a number sentence.

EUREKA MATH

Lesson 20:   Solve *take from with result unknown* expressions and equations using the minus sign with no unknown.

© 2018 Great Minds®. eureka-math.org

81

There were 4 oranges. Robin ate 1. Cross out the orange she ate. How many oranges were left? Fill in the boxes.

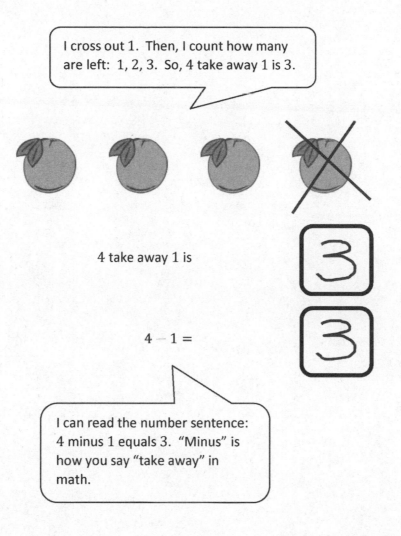

Lesson 21:   Represent subtraction story problems using objects, drawings, expressions, and equations.

© 2018 Great Minds®. eureka-math.org

83

Name _____   Date _____

There were 5 apples. Bill ate 1. Cross out the apple he ate. How many apples were left? Fill in the boxes.

5 take away 1 is ☐

5 - 1 = ☐

There were 5 oranges. Pat took 2. Draw the oranges. Cross out the 2 she took. How many oranges were left? Fill in the boxes.

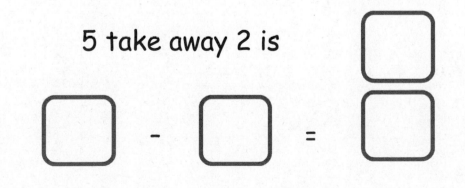

5 take away 2 is ☐

☐ - ☐ = ☐

Lesson 21:   Represent subtraction story problems using objects, drawings, expressions, and equations.

© 2018 Great Minds®. eureka-math.org

85

Draw 6 hearts. Cross out 2. Fill in the number sentence and the number bond.

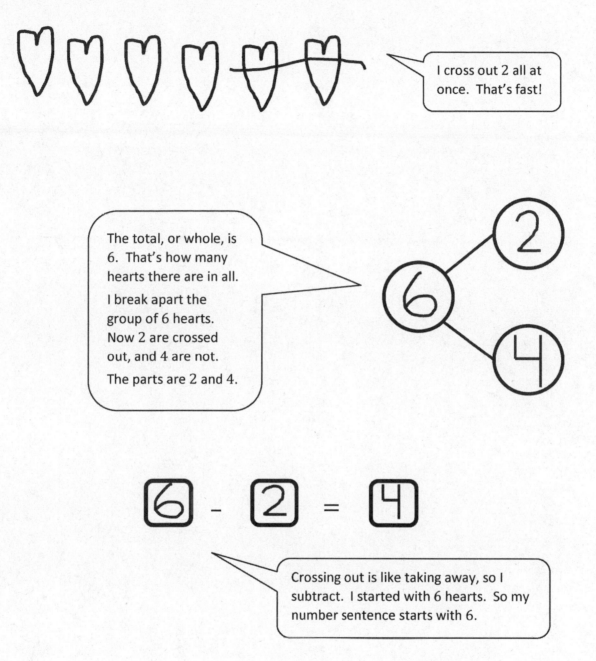

I cross out 2 all at once. That's fast!

The total, or whole, is 6. That's how many hearts there are in all.

I break apart the group of 6 hearts. Now 2 are crossed out, and 4 are not.

The parts are 2 and 4.

Crossing out is like taking away, so I subtract. I started with 6 hearts. So my number sentence starts with 6.

Lesson 22:   Decompose the number 6 using 5-group drawings by breaking off or
                  removing a part, and record each decomposition with a drawing and
                  subtraction equation.
© 2018 Great Minds®. eureka-math.org

Name _____   Date _____

Here are 6 books. Cross out 2. How many are left? Fill in the number bond and the number sentence.

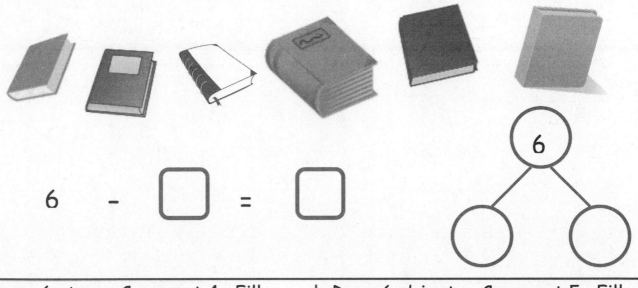

6 - ☐ = ☐

Draw 6 stars. Cross out 4. Fill in the number sentence and the number bond.	Draw 6 objects. Cross out 5. Fill in the number sentence and the number bond.
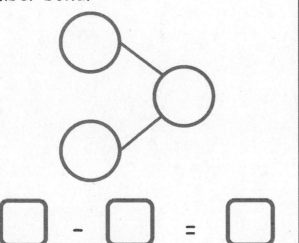	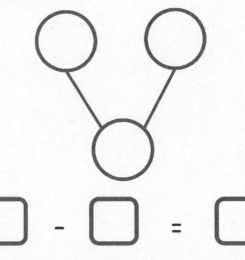
☐ - ☐ = ☐	☐ - ☐ = ☐

On the back of your paper, draw 6 triangles. Cross out 1. Write a number sentence, and draw a number bond to match.

EUREKA MATH

Lesson 22:   Decompose the number 6 using 5-group drawings by breaking off or removing a part, and record each decomposition with a drawing and subtraction equation.

© 2018 Great Minds®. eureka-math.org

Draw 7 dots in a 5-group. Cross out 4 dots. Fill in the number sentence and number bond.

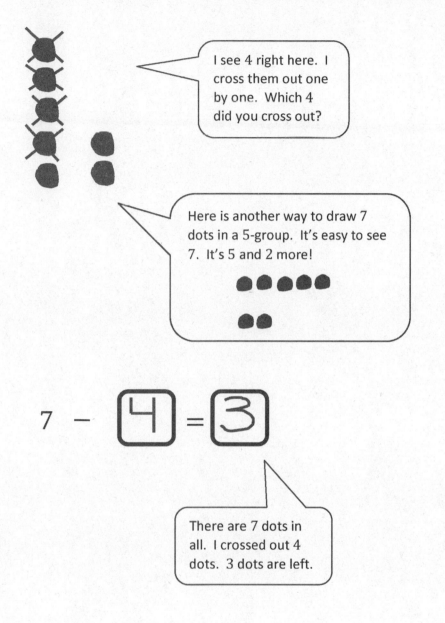

I see 4 right here. I cross them out one by one. Which 4 did you cross out?

4 dots are crossed out. 3 dots are not crossed out. 4 and 3 are the parts. 7 is the whole.

Here is another way to draw 7 dots in a 5-group. It's easy to see 7. It's 5 and 2 more!

$7 - 4 = 3$

There are 7 dots in all. I crossed out 4 dots. 3 dots are left.

Lesson 23: Decompose the number 7 using 5-group drawings by hiding a part, and record each decomposition with a drawing and subtraction equation.

91

© 2018 Great Minds®. eureka-math.org

Name _____   Date _____

Fill in the number sentence and number bond.
Cross out 5 dots.

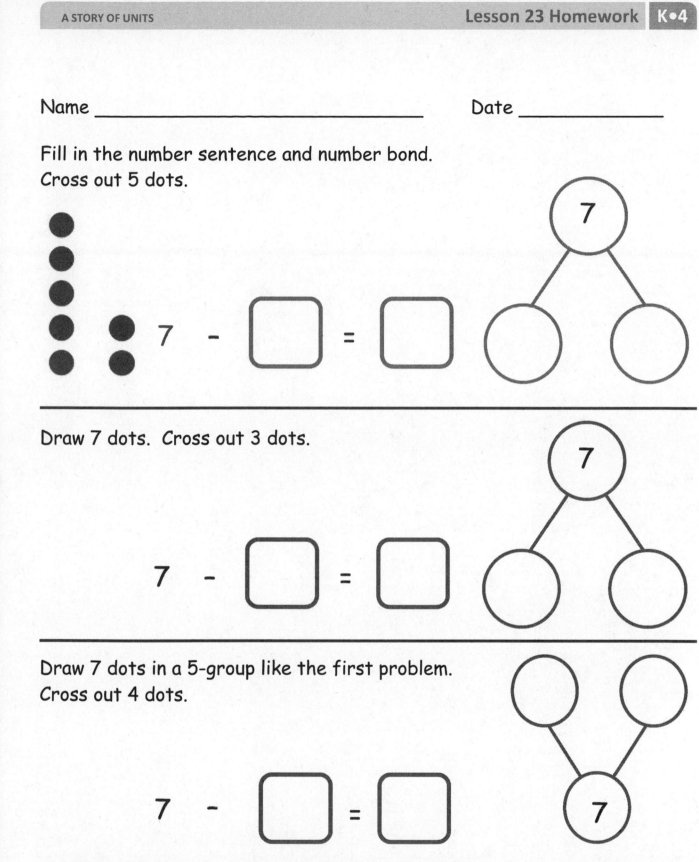

7 − ☐ = ☐   7

Draw 7 dots. Cross out 3 dots.

7 − ☐ = ☐   7

Draw 7 dots in a 5-group like the first problem.
Cross out 4 dots.

7 − ☐ = ☐   7

On the back of your paper, draw 7 dots. Cross out some, and write
a number sentence and a number bond to match.

EUREKA MATH

Lesson 23:   Decompose the number 7 using 5-group drawings by hiding a part, and
              record each decomposition with a drawing and subtraction equation.

© 2018 Great Minds®. eureka-math.org

93

Here is 8 the 5-group way. Put an X on 3 cubes. How many are left?

Fill in the number sentence and number bond.

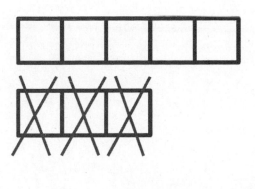

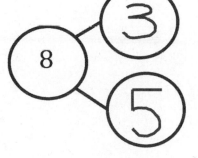

$8 - \boxed{3} = \boxed{5}$

> I did it! My picture, my number bond, and my number sentence all match. I could have crossed out 3 cubes a different way, and it would still match.

> My number bond shows how I broke apart 8. 3 cubes are crossed out. 5 cubes are not crossed out. 8 is the total, or whole. It's like 3 and 5 are hiding inside of 8.

Name _____   Date _____

Here is 8 the 5-group way. Put an X on 2 cubes. How many are left?
Fill in the number sentence and number bond.

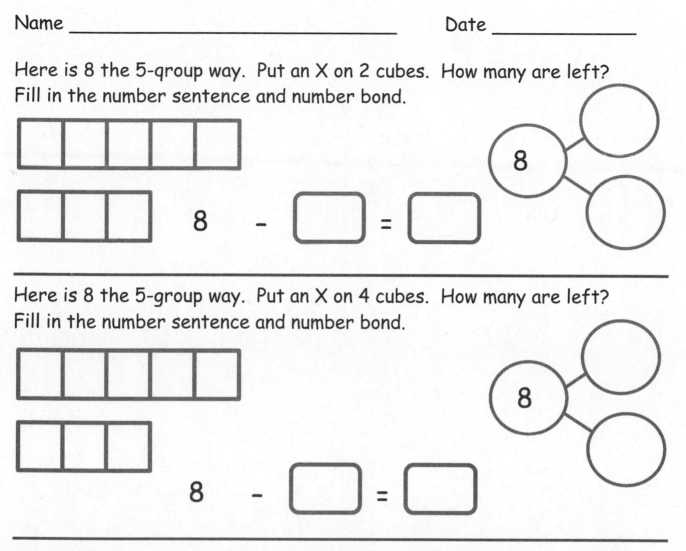

Here is 8 the 5-group way. Put an X on 4 cubes. How many are left?
Fill in the number sentence and number bond.

Draw 8 the 5-group way. Put an X on some cubes. How many are left?
Write the number sentence and the number bond.

On the back of your paper, draw 7 the 5-group way. Put an X on some, and
write a number sentence and number bond.

There are 9 stars. Color some blue and the rest yellow. Fill in the number bond.

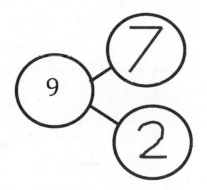

This is fun! I get to choose how many of each color. There are so many ways to break apart 9.

Let me tell you how my number bond goes with my star picture. There are 9 stars in all. That's the total. I color 7 blue and 2 yellow. Those are the parts. When I count all the stars, there are still 9.

Lesson 25: Model decompositions of 9 using a story situation, objects, and number bonds.

99

© 2018 Great Minds®. eureka-math.org

Name _____   Date _____

There are 9 leaves. Color some of them red and the rest of them yellow.
Fill in the number bond to match.

There are 9 acorns. Color some of them green and the rest yellow. Fill
in the number bond to match.

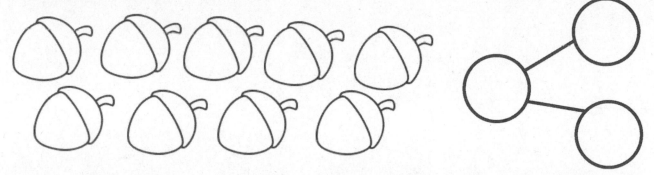

Draw 9 birds. Color some of them blue and the rest red. Fill in the
number bond to match.

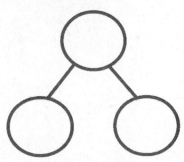

On the back of your paper, draw 9 triangles. Color some red and some
brown. Draw and fill in a number bond to match.

EUREKA MATH

Lesson 25:   Model decompositions of 9 using a story situation, objects, and
             number bonds.

101

© 2018 Great Minds®. eureka-math.org

The squares below represent cube sticks.
Do the linking cube sticks match the number bond? Circle yes or no.

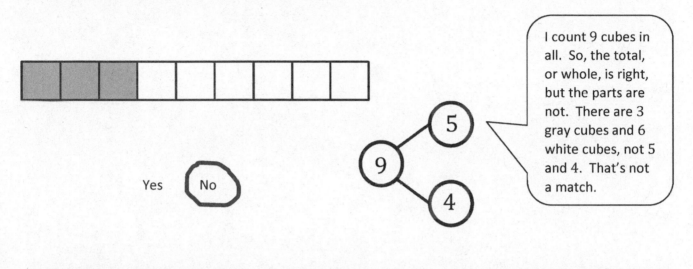

Yes   No

I count 9 cubes in all. So, the total, or whole, is right, but the parts are not. There are 3 gray cubes and 6 white cubes, not 5 and 4. That's not a match.

Make the number bond match the cube stick.

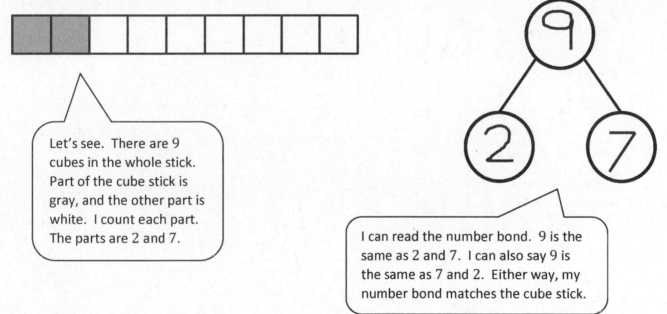

Let's see. There are 9 cubes in the whole stick. Part of the cube stick is gray, and the other part is white. I count each part. The parts are 2 and 7.

I can read the number bond. 9 is the same as 2 and 7. I can also say 9 is the same as 7 and 2. Either way, my number bond matches the cube stick.

Lesson 26: Model decompositions of 9 using fingers, linking cubes, and number bonds.

© 2018 Great Minds®. eureka-math.org

103

Name _____   Date _____

The squares below represent cube sticks.
Do the linking cube sticks match the number bond? Circle yes or no.

Yes   No

9
1   8

Yes   No

5
9
4

Yes   No

3
9
6

Lesson 26:   Model decompositions of 9 using fingers, linking cubes, and number
bonds.

© 2018 Great Minds®. eureka-math.org

105

Make the number bond match the cube stick.

Lesson 26:   Model decompositions of 9 using fingers, linking cubes, and number
bonds.

© 2018 Great Minds®. eureka-math.org

Pretend this is your bracelet.

Color some beads red and the rest black. Make a number bond to match.

Cool! I get to choose how many of each color. I pick 7 red and 3 black. My friend might pick different numbers. No matter what, the total of number of beads on each of our bracelets is still 10.

R R R R R R R B B B

10 7 3

The whole, or total, is 10. The parts are 7 and 3. There are 10 beads on the whole entire bracelet. The number 7 is for just the red beads, and the number 3 is for just the black beads.

Lesson 27:   Model decompositions of 10 using a story situation, objects, and number bonds.

107

© 2018 Great Minds®. eureka-math.org

Name _____ Date _____

Pretend this is your bracelet.
Color 5 beads blue and the rest green. Make a number bond to match.

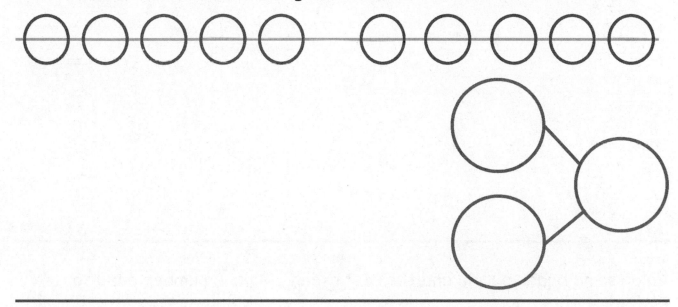

Color some beads yellow and the rest orange. Make a number bond to match.

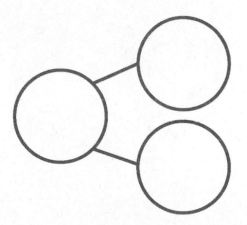

Lesson 27:   Model decompositions of 10 using a story situation, objects, and
              number bonds.

109

© 2018 Great Minds®. eureka-math.org

Color some beads yellow and the rest black. Make a number bond to match.

○──○──○──○──○──○──○──○──○──○

Color some beads purple and the rest green. Make a number bond to match.

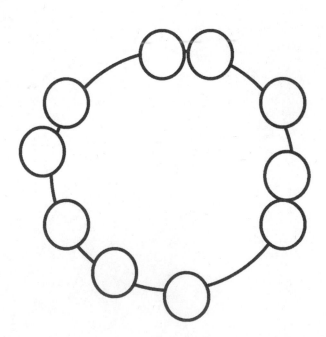

Lesson 27:   Model decompositions of 10 using a story situation, objects, and
                                number bonds.

© 2018 Great Minds®. eureka-math.org

EUREKA MATH

Write a number bond to match each domino.

It's easy to break apart numbers with dominoes. Just count the number of dots on each side to get the parts.

I count all the dots to find the total.

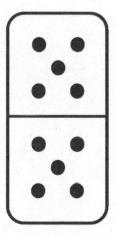

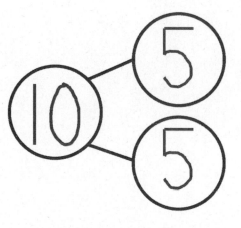

There are so many ways to break apart 10. This one is just like the fingers on both of my hands!

Lesson 28:   Model decompositions of 10 using fingers, sets, linking cubes, and number bonds.

© 2018 Great Minds®. eureka-math.org

111

Name _____   Date _____

Write a number bond to match each domino.

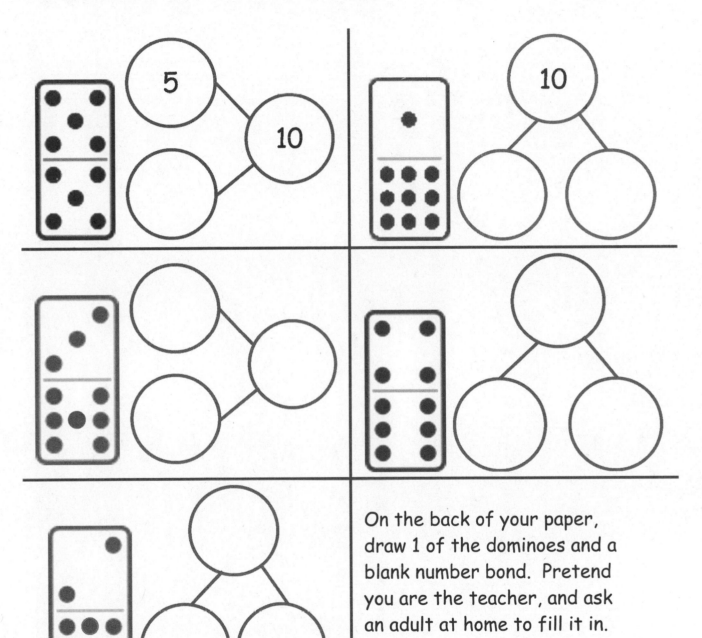

On the back of your paper, draw 1 of the dominoes and a blank number bond. Pretend you are the teacher, and ask an adult at home to fill it in.

Lesson 28:   Model decompositions of 10 using fingers, sets, linking cubes, and number bonds.

113

© 2018 Great Minds®. eureka-math.org

PURINA

Rosey found 8 paintbrushes and 1 gluestick. She found 9 art things. Draw the paintbrushes and the glue stick in the 5-group way. Fill in the number sentence.

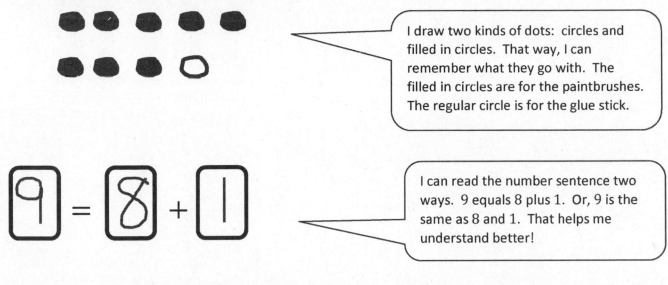

I draw two kinds of dots: circles and filled in circles. That way, I can remember what they go with. The filled in circles are for the paintbrushes. The regular circle is for the glue stick.

I can read the number sentence two ways. 9 equals 8 plus 1. Or, 9 is the same as 8 and 1. That helps me understand better!

Jack needs a snack. He found 9 pieces of fruit. 5 were apples, and 4 were oranges. Draw the apples and oranges in the 5-group way.
Fill in the number sentence.

To draw the 5-group way, I draw dots on the top row, from left to right. 9 is 5 and 4, so I draw 5 dots on the top, and 4 on the bottom.

Lesson 29:   Represent pictorial decomposition and composition addition stories to 9 with 5-group drawings and equations with no unknown.   115

© 2018 Great Minds®. eureka-math.org

Name _____   Date _____

Jack found 7 balls while cleaning the toy bin. He found 6 basketballs and 1 baseball. Fill in the number sentence and the number bond.

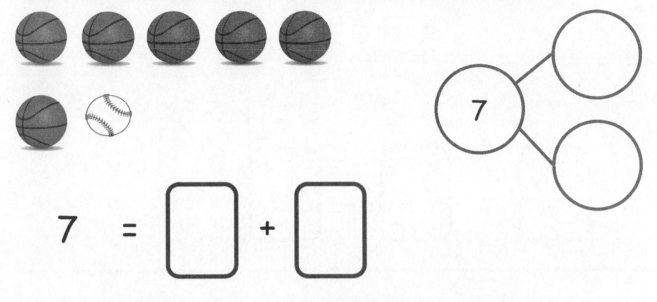

7 = ☐ + ☐

Jack found 7 mitts and 2 bats. He found 9 things. Fill in the number sentence and the number bond.

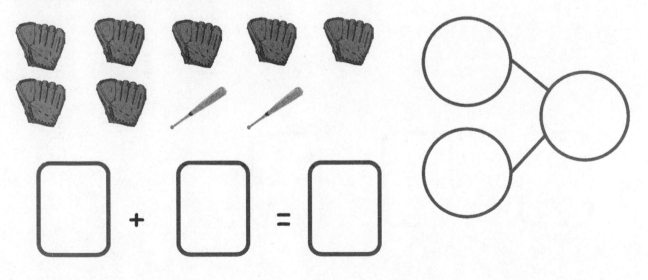

☐ + ☐ = ☐

Lesson 29:   Represent pictorial decomposition and composition addition stories to 9 with 5-group drawings and equations with no unknown.

© 2018 Great Minds®. eureka-math.org

117

Jack found 8 hockey pucks and 1 hockey stick. He found 9 hockey things. Draw the hockey pucks and stick in the 5-group way. Fill in the number sentence.

Jack needs a snack. He found 9 pieces of fruit. 5 were strawberries, and 4 were grapes. Draw the strawberries and grapes in the 5-group way. Fill in the number sentence.

Lesson 29:   Represent pictorial decomposition and composition addition stories to 9 with 5-group drawings and equations with no unknown.

© 2018 Great Minds®. eureka-math.org

EUREKA MATH

Ming saw 10 animals at the pet store. She saw 6 fish and 4 turtles. Draw the animals in the 5-group way.

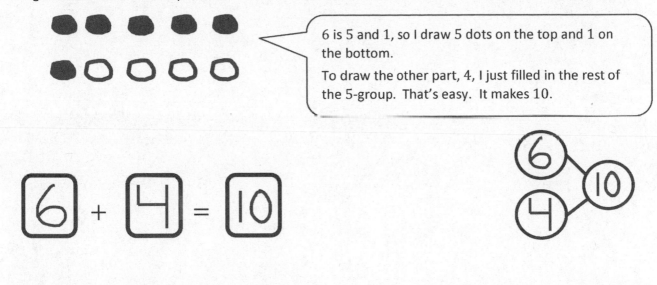

6 is 5 and 1, so I draw 5 dots on the top and 1 on the bottom.

To draw the other part, 4, I just filled in the rest of the 5-group. That's easy. It makes 10.

$6 + 4 = 10$

Make 2 groups. Circle 1 of the groups. Write a number sentence to match. Find as many partners of 10 as you can.

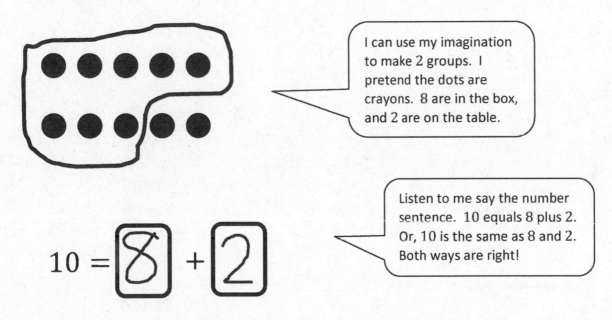

I can use my imagination to make 2 groups. I pretend the dots are crayons. 8 are in the box, and 2 are on the table.

$10 = 8 + 2$

Listen to me say the number sentence. 10 equals 8 plus 2. Or, 10 is the same as 8 and 2. Both ways are right!

Lesson 30: Represent pictorial decomposition and composition addition stories to 10 with 5 group drawings and equations with no unknown.

119

© 2018 Great Minds®. eureka-math.org

Name _____   Date _____

Fill in the number bonds, and complete the number sentences.

Scott went to the zoo. He saw 6 giraffes and 4 zebras. He saw 10 animals altogether.

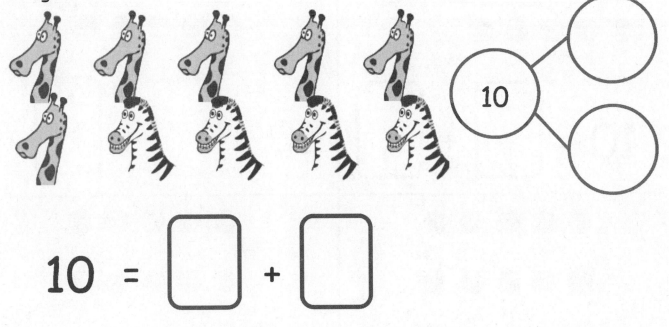

10 = ⬚ + ⬚

Susan saw 10 animals at the zoo. She saw 5 lions and 5 elephants. Draw the animals in the 5-group way.

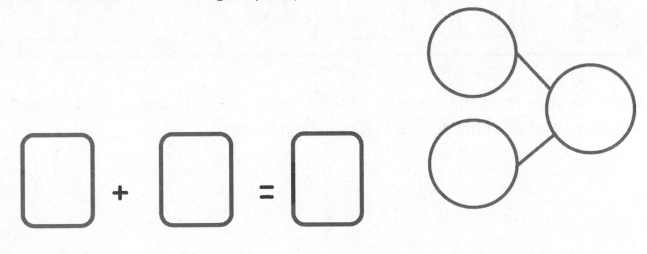

⬚ + ⬚ = ⬚

EUREKA MATH

Lesson 30:   Represent pictorial decomposition and composition addition
             stories to 10 with 5 group drawings and equations with no unknown.

121

© 2018 Great Minds®. eureka-math.org

Make 2 groups. Circle 1 of the groups. Write a number sentence to match.
Find as many partners of 10 as you can.

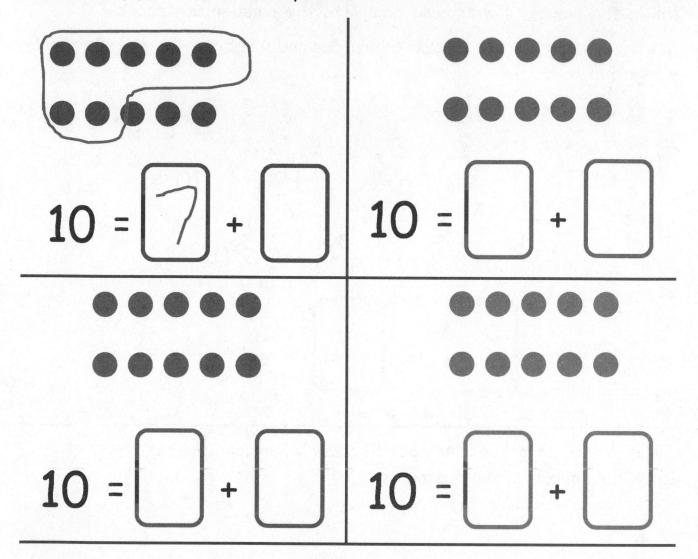

Draw 10 dots the 5-group way. Make 2 groups. Circle one of the groups.
Write a number sentence to match your drawing.

Lesson 30:   Represent pictorial decomposition and composition addition
                    stories to 10 with 5 group drawings and equations with no unknown.

© 2018 Great Minds®. eureka-math.org

Draw the story. Fill in the number sentence.

Ke'Azia has 6 chocolate chip cookies and 3 sugar cookies. How many cookies does she have altogether?

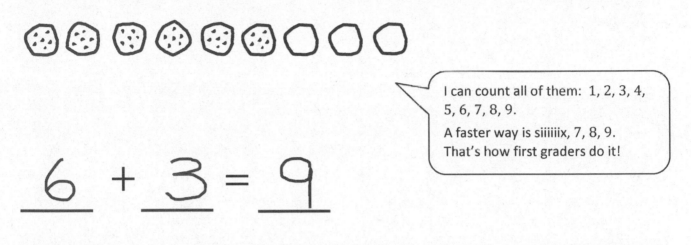

> I can count all of them: 1, 2, 3, 4, 5, 6, 7, 8, 9.
>
> A faster way is siiiiiix, 7, 8, 9. That's how first graders do it!

$$\underline{6} + \underline{3} = \underline{9}$$

Mario's mother bought juice boxes. 5 were lemonade, and 4 were fruit punch. How many juice boxes did she have in all?

> Math drawings don't have to look like the real thing. I can just put an L, and my teacher will know it's lemonade.

$$\underline{5} + \underline{4} = \underline{9}$$

EUREKA MATH

Lesson 31: Solve *add to with total unknown* and *put together with total unknown* problems with totals of 9 and 10.

123

© 2018 Great Minds®. eureka-math.org

Name _____   Date _____

Draw the story.   Fill in the number sentence.

Jake has 7 chocolate cookies and 2 sugar cookies. How many cookies does he have altogether?

_____ + _____ = _____

Jake's mother bought juice boxes. 4 were apple juice, and 5 were orange juice. How many juice boxes did she have in all?

_____ + _____ = _____

EUREKA MATH   Lesson 31:   Solve *add to with total unknown* and *put together with total*   125
                                  unknown problems with totals of 9 and 10.

© 2018 Great Minds®. eureka-math.org

Draw the story.   Write a number sentence to match.

Ryan had 5 celery sticks and 5 carrot sticks. How many veggie sticks did Ryan have altogether?

Draw an addition story, and write a number sentence to match it. Explain your work to an adult at home.

Lesson 31:   Solve *add to with total unknown* and *put together with total unknown* problems with totals of 9 and 10.

© 2018 Great Minds®. eureka-math.org

Anya has 9 stuffed cats. Some are gray, and the rest are white. Show two different ways Anya's cats could look. Fill in the number sentences to match.

I colored this one the 5-group way.

I colored this one a different way.

$$9 = \boxed{5} + \boxed{4}$$

$$9 = \boxed{6} + \boxed{3}$$

9 is the same as 5 and 4. It is also the same as 6 and 3. There is more than one way to break apart 9.

Lesson 32:   Solve *both addends unknown* word problems with totals of 9 and 10 using 5-group drawings.

127

© 2018 Great Minds®. eureka-math.org

Name _____   Date _____

Jerry has 9 baseball hats. Draw the hats the 5-group way. Color some red and some blue. Fill in the number sentence to match.

$$9 = \boxed{} + \boxed{}$$

_____

Anne had 10 pencils. Draw the pencils the 5-group way. Color some pencils blue and some yellow. Fill in the number sentence to match.

$$10 = \boxed{} + \boxed{}$$

EUREKA MATH

Lesson 32:   Solve *both addends unknown* word problems with totals of 9 and 10 using 5-group drawings.

129

© 2018 Great Minds®. eureka-math.org

There are 10 apples. Color some red and the rest green. Then, show a different way the apples could look. Fill in the number sentences to match.

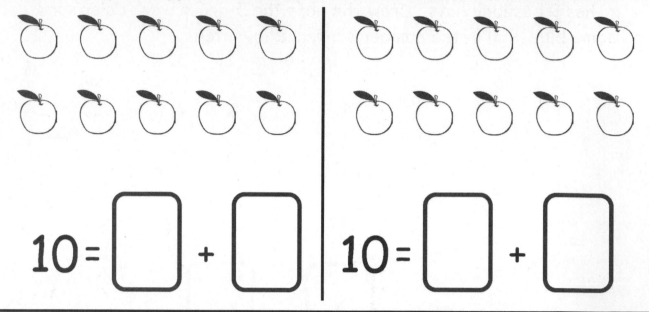

10 = ☐ + ☐   10 = ☐ + ☐

Anya has 9 stuffed cats. Some are orange, and the rest are gray. Show two different ways Anya's cats could look. Fill in the number sentences to match.

9 = ☐ + ☐   9 = ☐ + ☐

Lesson 32:   Solve *both addends unknown* word problems with totals of 9 and 10 using 5-group drawings.

© 2018 Great Minds®. eureka-math.org

Fill in the number sentence to match the story.

There were 10 teddy bears. Cross out 3 bears. There are 7 bears left.

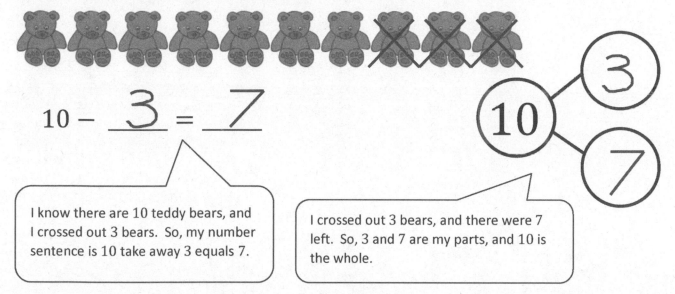

$$10 - \underline{\ 3\ } = \underline{\ 7\ }$$

I know there are 10 teddy bears, and I crossed out 3 bears. So, my number sentence is 10 take away 3 equals 7.

I crossed out 3 bears, and there were 7 left. So, 3 and 7 are my parts, and 10 is the whole.

Draw a line from the picture to the number sentence it matches.

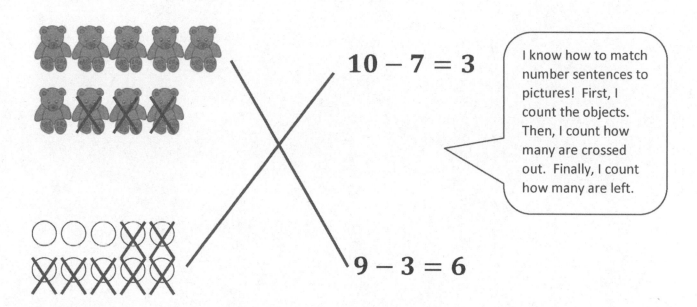

$$10 - 7 = 3$$

$$9 - 3 = 6$$

I know how to match number sentences to pictures! First, I count the objects. Then, I count how many are crossed out. Finally, I count how many are left.

© 2018 Great Minds®. eureka-math.org

Name _____   Date _____

Fill in the number sentence and the number bond.

There were 10 teddy bears. Cross out 2 bears. There are 8 bears left.

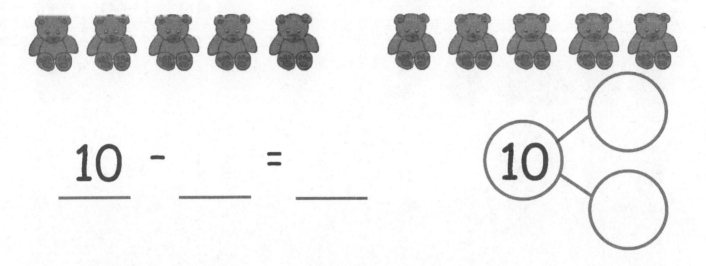

10 - ____ = ____

There were 10 teddy bears. Cross out 9. There is 1 left.

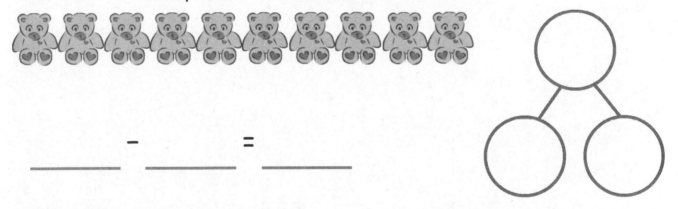

____ - ____ = ____

There were 10 teddy bears. Cross out 3. There are 7 bears left.

____ - ____ = ____

EUREKA
MATH

Lesson 33:   Solve *take from* equations with no unknown using numbers to 10.

133

© 2018 Great Minds®. eureka-math.org

Draw a line from the picture to the number sentence it matches.

   10 − 1 = 9

   10 − 3 = 7

   9 − 4 = 5

9 − 8 = 1

© 2018 Great Minds®. eureka-math.org

There were 10 penguins. 4 penguins went back to the ship. Cross out 4 penguins. Fill in the number sentence and the number bond.

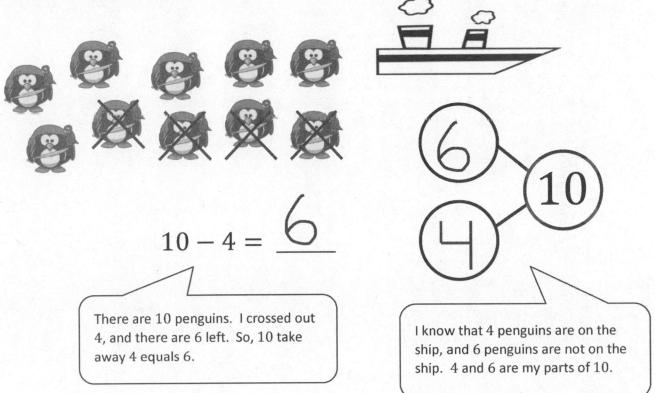

$$10 - 4 = \underline{6}$$

There are 10 penguins. I crossed out 4, and there are 6 left. So, 10 take away 4 equals 6.

I know that 4 penguins are on the ship, and 6 penguins are not on the ship. 4 and 6 are my parts of 10.

The squares below represent cubes. Count the cubes. Draw a line to break 4 cubes off the train. Fill in the number sentence and the number bond.

I drew my line to break apart my cube train into parts of 4 and 2. I have 6 cubes. I break off 4 cubes, and I have 2 cubes left!

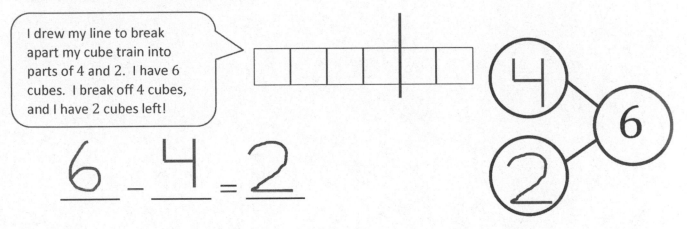

$$\underline{6} - \underline{4} = 2$$

Lesson 34:   Represent subtraction story problems by breaking off, crossing out, and hiding a part.

135

© 2018 Great Minds®. eureka-math.org

Name _____   Date _____

There were 8 penguins. 2 penguins went back to the ship. Cross out 2 penguins. Fill in the number sentence and the number bond.

8 - 2 = ___ ___

The squares below represent cubes.
Count the cubes. Draw a line to break 4 cubes off the train. Fill in the number sentence and the number bond.

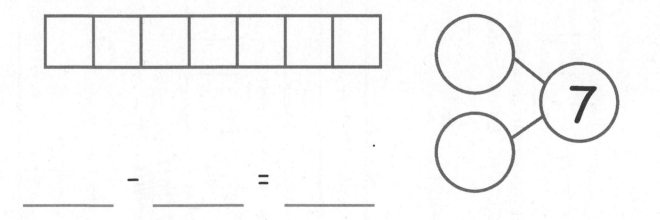

___ - ___ = ___

EUREKA MATH

Lesson 34:   Represent subtraction story problems by breaking off, crossing out, and hiding a part.

137

© 2018 Great Minds®. eureka-math.org

There are 10 bears. Some go inside the cave to hide. Cross them out.
Complete the number sentence.

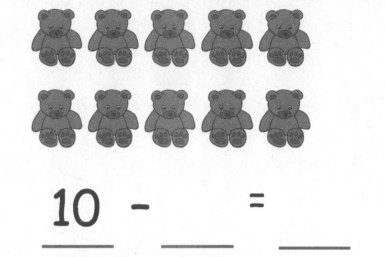

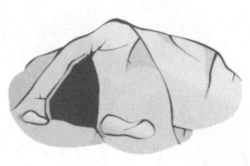

10 - _____ = _____

____

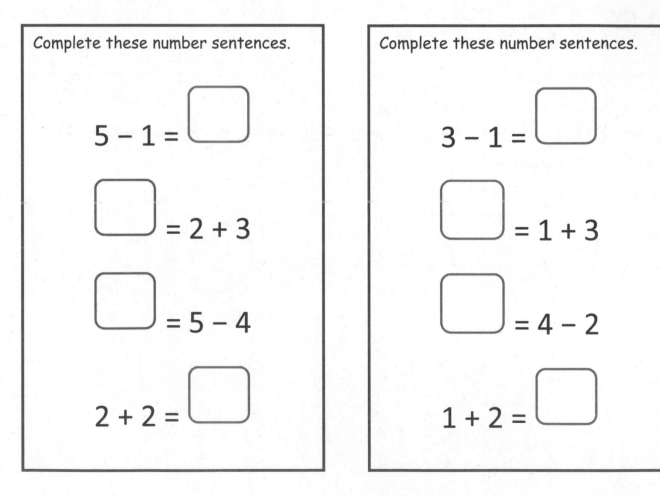

Complete these number sentences.

5 – 1 = ☐

☐ = 2 + 3

☐ = 5 – 4

2 + 2 = ☐

Complete these number sentences.

3 – 1 = ☐

☐ = 1 + 3

☐ = 4 – 2

1 + 2 = ☐

Lesson 34: Represent subtraction story problems by breaking off, crossing out, and hiding a part.

© 2018 Great Minds®. eureka-math.org

Cross off the part that goes away. Fill in the number bond and number sentence.

Mary had 9 library books. She returned 2 books to the library. How many books are left?

I solved it! If Mary had 9 library books, and she returns 2, then she has 7 books left.

$$9 - 2 = 7$$

Make a 5-group drawing to show the story. Cross off the part that goes away. Fill in the number bond and number sentence.

Ryder had 9 pencils. 4 of them broke. How many pencils are left?

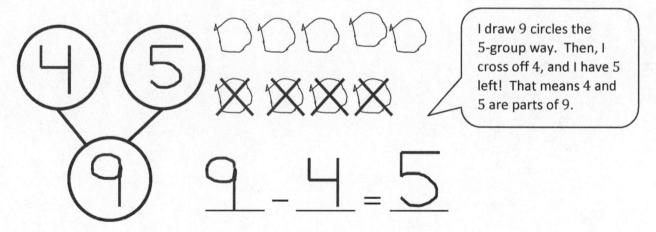

I draw 9 circles the 5-group way. Then, I cross off 4, and I have 5 left! That means 4 and 5 are parts of 9.

$$9 - 4 = 5$$

Lesson 35:   Decompose the number 9 using 5-group drawings, and record each decomposition with a subtraction equation.

139

© 2018 Great Minds®. eureka-math.org

Name _____   Date _____

Cross off the part that goes away. Fill in the number bond and number sentence.

Mary had 9 library books. She returned 1 book to the library. How many books are left?

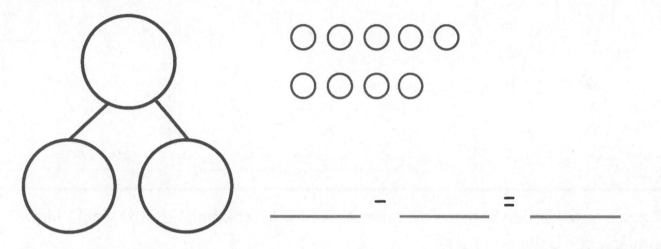

_____ - _____ = _____

There were 9 lunch bags. 3 bags were thrown away. How many bags are there now?

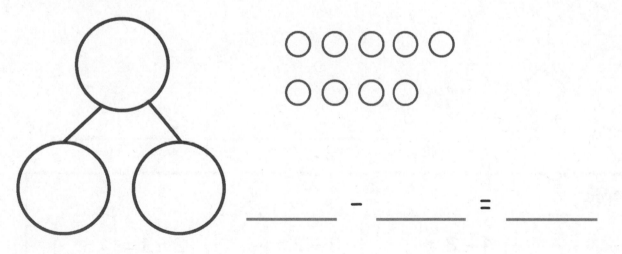

_____ - _____ = _____

Lesson 35:   Decompose the number 9 using 5-group drawings, and record each decomposition with a subtraction equation.

141

© 2018 Great Minds®. eureka-math.org

Make a 5-group drawing to show the story. Cross off the part that goes away. Fill in the number bond and number sentence.

Ms. Lopez has 9 pencils. 7 of them broke. How many pencils are left?

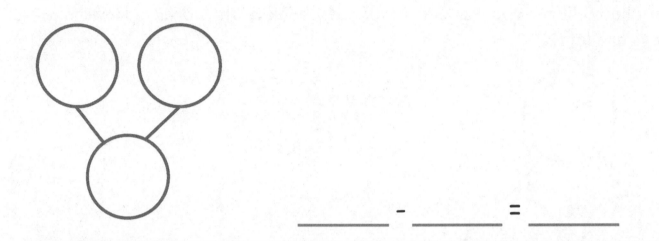

_____ - _____ = _____

There are 9 soccer balls. The team kicked 5 of the balls at the goal. How many soccer balls are left?

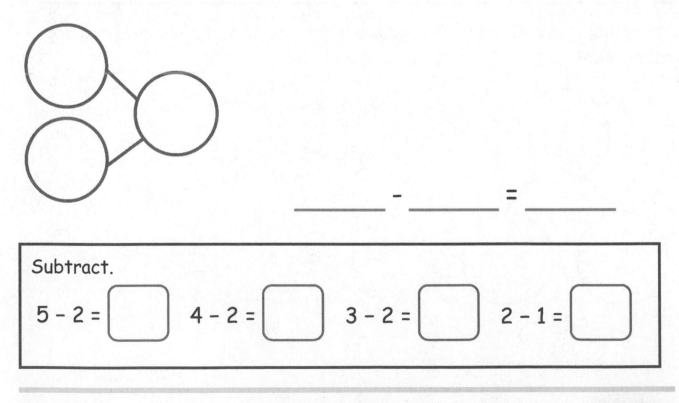

_____ - _____ = _____

Subtract.

5 – 2 = ☐   4 – 2 = ☐   3 – 2 = ☐   2 – 1 = ☐

Lesson 35:   Decompose the number 9 using 5-group drawings, and record each decomposition with a subtraction equation.

© 2018 Great Minds®. eureka-math.org

Fill in the number bond and number sentence. Cross off the part that goes away.

MacKenzie had 10 buttons on her jacket. 4 buttons broke off her jacket. How many buttons are left on her jacket?

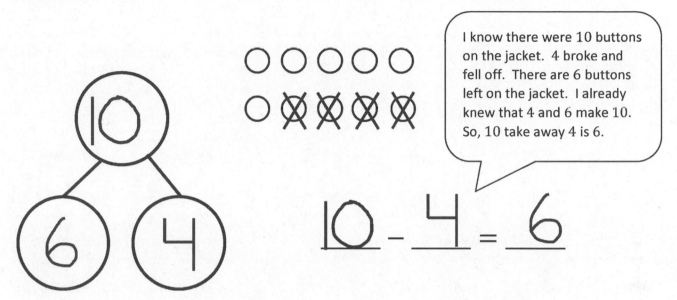

I know there were 10 buttons on the jacket. 4 broke and fell off. There are 6 buttons left on the jacket. I already knew that 4 and 6 make 10. So, 10 take away 4 is 6.

$$10 - 4 = 6$$

Make a 5-group drawing to show the story. Fill in the number bond and number sentence. Cross off the part that goes away.

Bob had 10 toy cars. 3 cars drove away. How many cars are left?

I made a 5-group drawing to show the cars. 3 drove away, so I crossed out 3. There are 7 cars left.

$$10 - 3 = 7$$

Lesson 36:   Decompose the number 10 using 5-group drawings, and record each decomposition with a subtraction equation.   143

© 2018 Great Minds®. eureka-math.org

Name _____   Date _____

Fill in the number bond and number sentence. Cross off the part that goes away.

MacKenzie had 10 buttons on her jacket. 2 buttons broke off her jacket. How many buttons are left on her jacket?

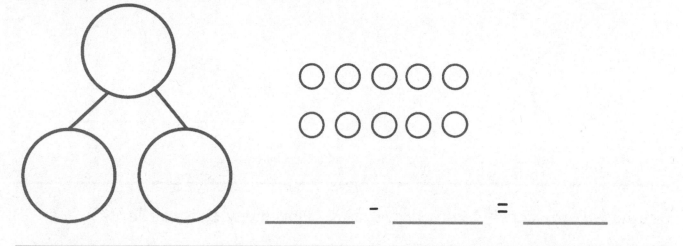

_____ - _____ = _____

Donna had 10 cups. 6 cups fell and broke. How many unbroken cups are there now?

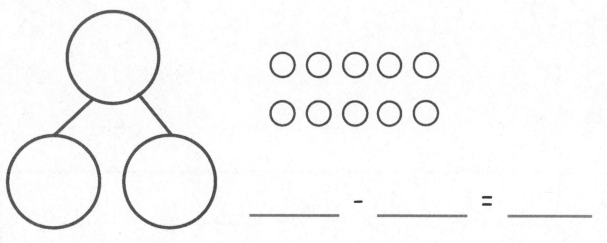

_____ - _____ = _____

EUREKA MATH

Lesson 36:   Decompose the number 10 using 5-group drawings, and record
each decomposition with a subtraction equation.

© 2018 Great Minds®. eureka-math.org

145

Make a 5-group drawing to show the story. Fill in the number bond and number sentence. Cross off the part that goes away.

There were 10 butterflies. 9 butterflies flew away. How many are left?

_____ - _____ = _____

Bob had 10 toy cars. 4 cars drove away. How many cars are left?

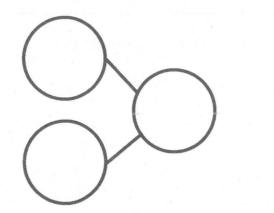

_____ - _____ = _____

Subtract.

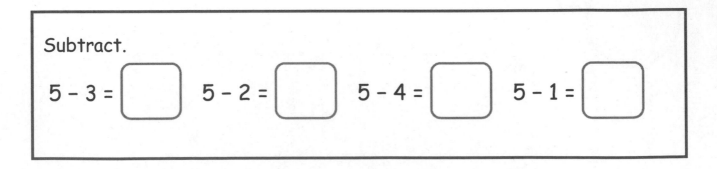

5 − 3 = ☐   5 − 2 = ☐   5 − 4 = ☐   5 − 1 = ☐

Lesson 36:   Decompose the number 10 using 5-group drawings, and record each decomposition with a subtraction equation.

© 2018 Great Minds®. eureka-math.org

EUREKA MATH

Listen to each story. Show the story with your fingers on the number path. Then, fill in the number sentence and number bond.

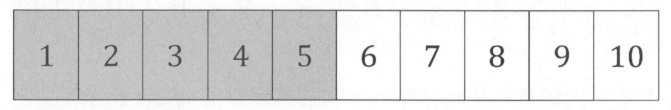

Joey had 7 pennies. He found 2 pennies in the couch. How many pennies does Joey have now?

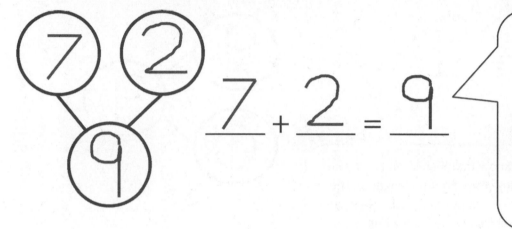

$$\underline{7} + \underline{2} = \underline{9}$$

I use the number path to help me solve the problem! I put my finger on 7 because Joey had 7 pennies. He found 2 pennies, so I hop forward 2 on the number path. My fingers stop on the 9. Joey has 9 pennies!

Joey gave the 2 pennies to his dad. How many pennies does Joey have now?

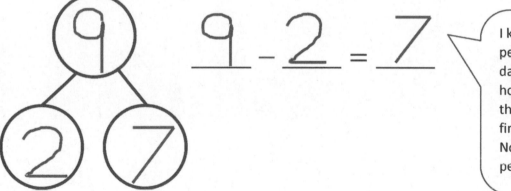

$$\underline{9} - \underline{2} = \underline{7}$$

I know that Joey has 9 pennies. He gave his dad 2 pennies, so I hop 2 backward on the number path. My fingers stop on the 7. Now, Joey has 7 pennies!

EUREKA
MATH

© 2018 Great Minds®. eureka-math.org

| 1 | 2 | 3 | 4 | 5 | 6 | 7 | 8 | 9 | 10 |

There were 9 children waiting for the school bus. No more children came to the bus stop. How many children are waiting now?

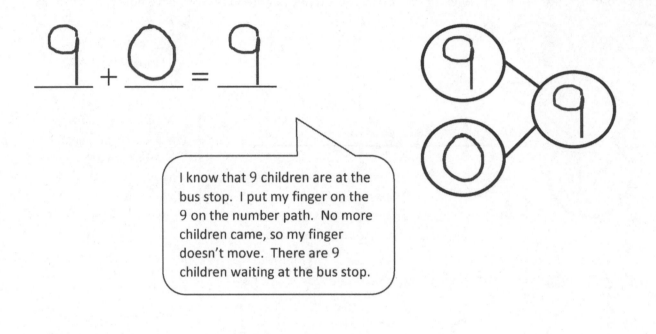

> I know that 9 children are at the bus stop. I put my finger on the 9 on the number path. No more children came, so my finger doesn't move. There are 9 children waiting at the bus stop.

Lesson 37:   Add or subtract 0 to get the same number and relate to word problems wherein the same quantity that joins a set, separates.

© 2018 Great Minds®. eureka-math.org

Name _____   Date _____

Listen to each story. Show the story with your fingers on the number path. Then, fill in the number sentence and number bond.

1	2	3	4	5	6	7	8	9	10

Joey had 5 pennies. He found 3 pennies in the couch. How many pennies does Joey have now?

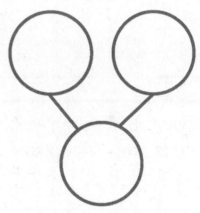

_____ + _____ = _____

Joey gave the 3 pennies to his dad. How many pennies does Joey have now?

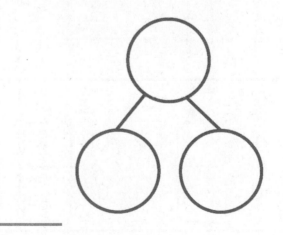

_____ - _____ = _____

Lesson 37:   Add or subtract 0 to get the same number and relate to word problems wherein the same quantity that joins a set, separates.

© 2018 Great Minds®. eureka-math.org

149

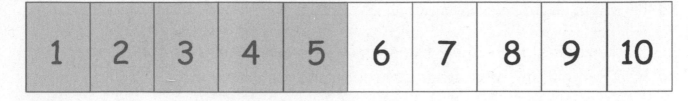

| 1 | 2 | 3 | 4 | 5 | 6 | 7 | 8 | 9 | 10 |

Siri had 9 pennies. She looked all around the house but could not find any more pennies. How many pennies does she have now?

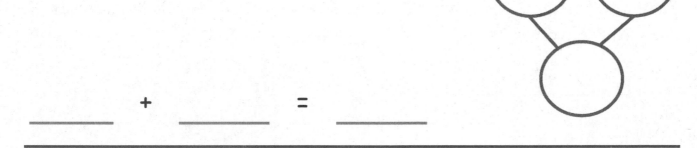

_____ + _____ = _____

There were 8 children waiting for the school bus. No more children came to the bus stop. How many children are waiting now?

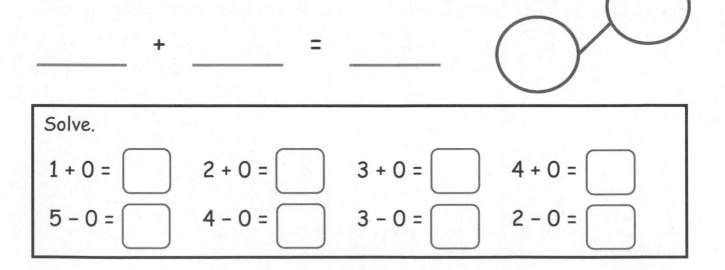

_____ + _____ = _____

Solve.

1 + 0 = ☐   2 + 0 = ☐   3 + 0 = ☐   4 + 0 = ☐

5 – 0 = ☐   4 – 0 = ☐   3 – 0 = ☐   2 – 0 = ☐

Lesson 37:   Add or subtract 0 to get the same number and relate to word problems wherein the same quantity that joins a set, separates.

© 2018 Great Minds®. eureka-math.org

Follow the instructions to color the 5-group. Then, fill in the number sentence and number bond to match.

Color 6 squares green and 1 square blue.

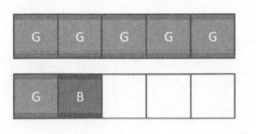

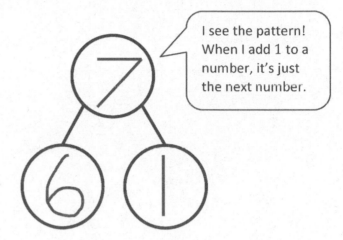

I see the pattern! When I add 1 to a number, it's just the next number.

$6 + 1 = 7$

Color 3 squares green and 1 square blue.

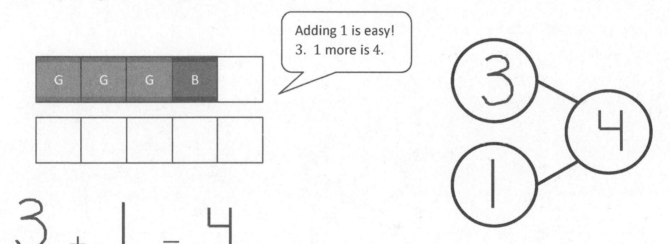

Adding 1 is easy!
3. 1 more is 4.

$3 + 1 = 4$

Lesson 38: Add 1 to numbers 1–9 to see the pattern of *the next number* using 5-group drawings and equations.

151

© 2018 Great Minds®. eureka-math.org

Name _____   Date _____

Follow the instructions to color the 5-group. Then, fill in the number sentence or number bond to match.

Color 9 squares green and 1 square blue.

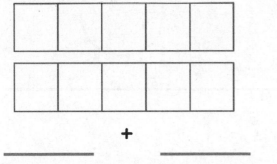

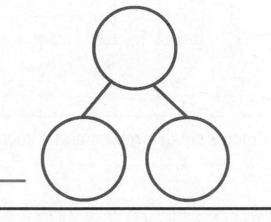

_____ + _____ = _____

Color 8 squares green and 1 square blue.

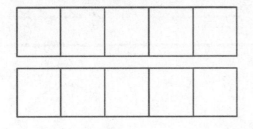

_____ + _____ = _____

Color 7 squares green and 1 square blue.

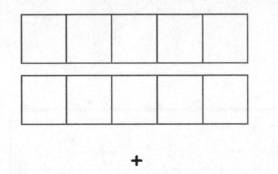

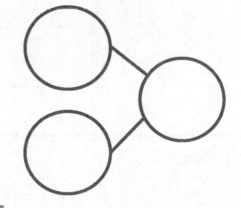

_____ + _____ = _____

Lesson 38:   Add 1 to numbers 1–9 to see the pattern of *the next number* using
5-group drawings and equations.

153

© 2018 Great Minds®. eureka-math.org

Color 2 squares green and 1 square blue.

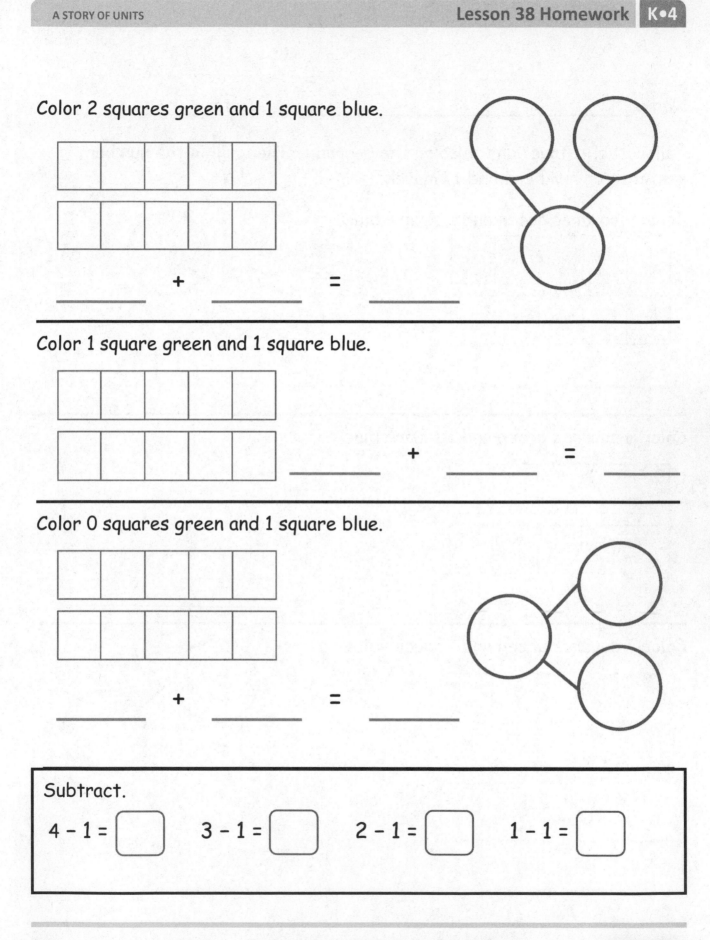

_____ + _____ = _____

Color 1 square green and 1 square blue.

_____ + _____ = _____

Color 0 squares green and 1 square blue.

_____ + _____ = _____

Subtract.

4 – 1 = ☐   3 – 1 = ☐   2 – 1 = ☐   1 – 1 = ☐

Lesson 38:   Add 1 to numbers 1–9 to see the pattern of *the next number* using 5-group drawings and equations.

© 2018 Great Minds®. eureka-math.org

EUREKA MATH®

Draw dots to make 10. Finish the number bonds. Draw a line from the 5-group to the matching number bond.

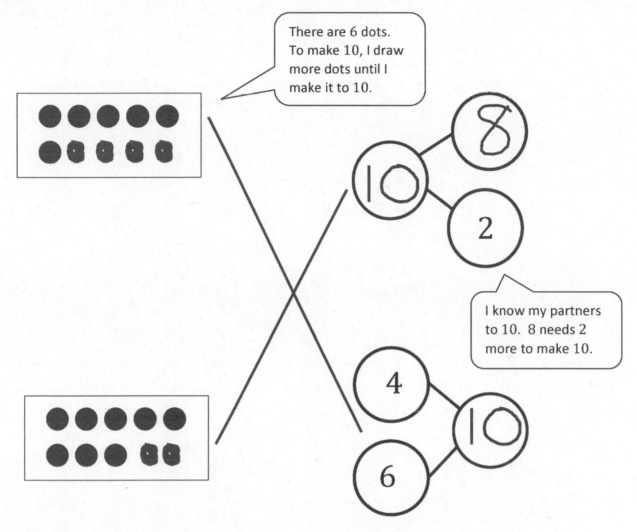

EUREKA MATH

Lesson 39:   Find the number that makes 10 for numbers 1–9, and record
               each with a 5-group drawing.

155

© 2018 Great Minds®. eureka-math.org

Name _____ Date _____

Draw dots to make 10. Finish the number bonds. Draw a line from the 5-group to the matching number bond.

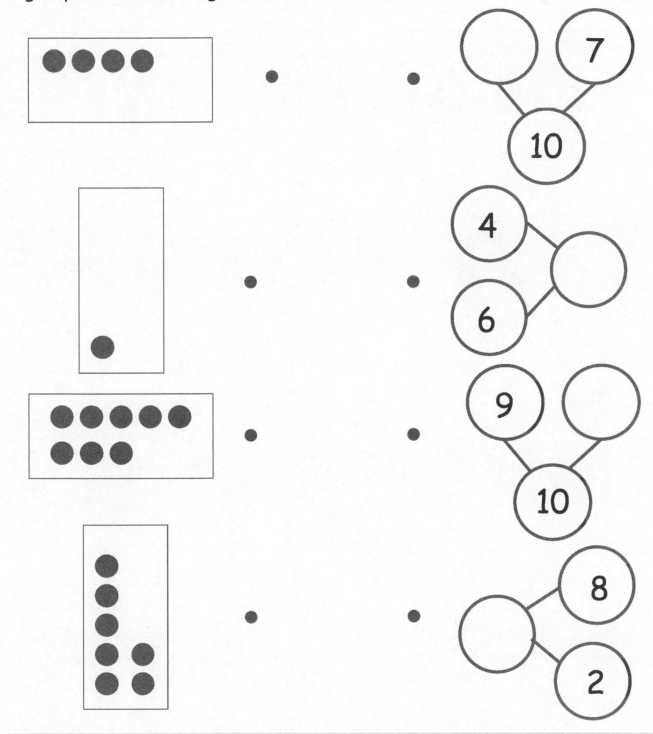

Lesson 39:   Find the number that makes 10 for numbers 1–9, and record
each with a 5-group drawing.

© 2018 Great Minds®. eureka-math.org

Color 7 boxes red the 5-group way. Color the rest blue to make 10. Fill in the number sentence.

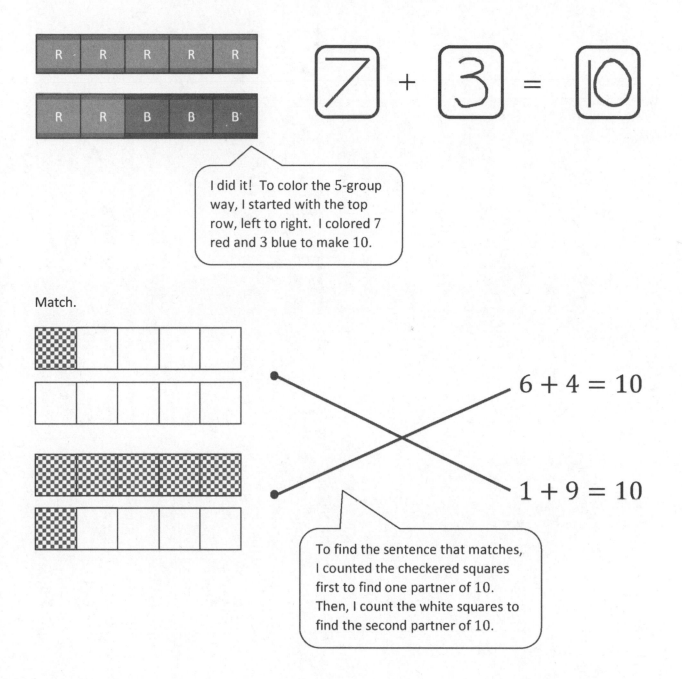

I did it! To color the 5-group way, I started with the top row, left to right. I colored 7 red and 3 blue to make 10.

Match.

$6 + 4 = 10$

$1 + 9 = 10$

To find the sentence that matches, I counted the checkered squares first to find one partner of 10. Then, I count the white squares to find the second partner of 10.

Lesson 40: Find the number that makes 10 for numbers 1–9, and record each with an addition equation.

159

© 2018 Great Minds®. eureka-math.org

Name _____   Date _____

Color 2 boxes red the 5-group way. Color the rest blue to make 10. Fill in the number sentence.

Color 5 boxes red the 5-group way. Color the rest blue to make 10. Fill in the number sentence.

Color 7 boxes red the 5-group way. Color the rest blue to make 10. Fill in the number sentence.

EUREKA MATH

Lesson 40:   Find the number that makes 10 for numbers 1–9, and record each with an addition equation.

161

© 2018 Great Minds®. eureka-math.org

Match.

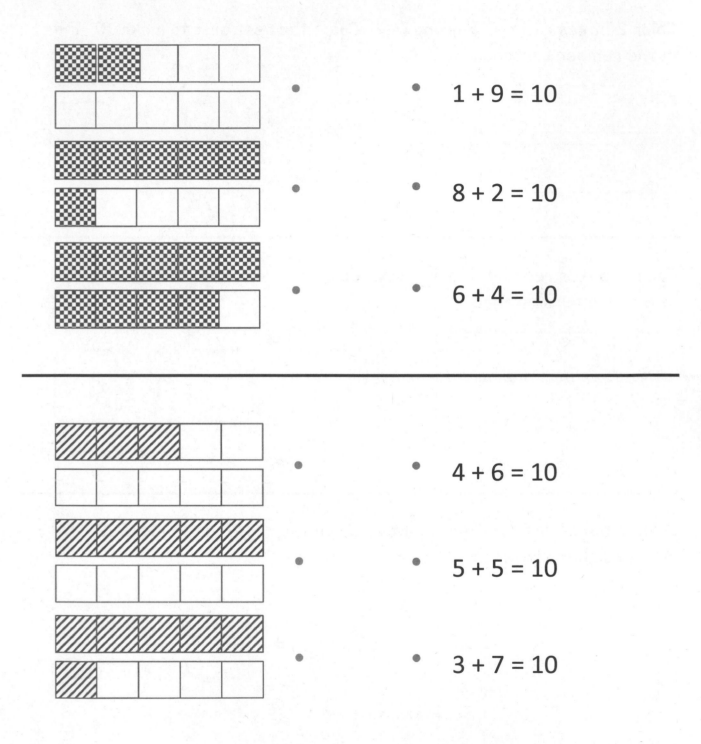

1 + 9 = 10

8 + 2 = 10

6 + 4 = 10

4 + 6 = 10

5 + 5 = 10

3 + 7 = 10

Lesson 40:   Find the number that makes 10 for numbers 1–9, and record each with an addition equation.

© 2018 Great Minds®. eureka-math.org

Complete a number bond and a number sentence for the problem:

Color some blocks orange and the rest yellow to make 10. All of the yellow blocks fell off the table. How many blocks are left?

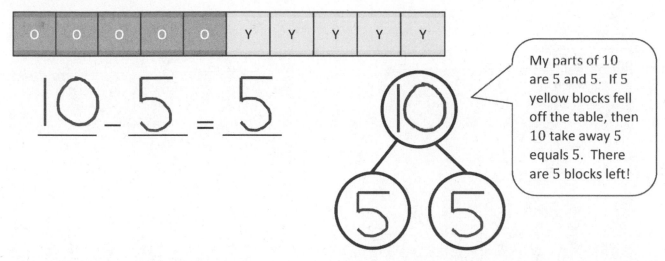

$$10 - 5 = 5$$

My parts of 10 are 5 and 5. If 5 yellow blocks fell off the table, then 10 take away 5 equals 5. There are 5 blocks left!

There were 10 horses in the yard. Some were brown, and some were white. Draw the horses the 5-group way. The brown ones went back into the barn. How many horses were still in the yard? Draw a number bond, and write a subtraction sentence.

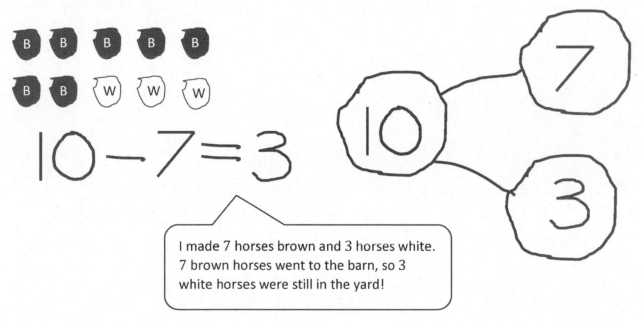

$$10 - 7 = 3$$

I made 7 horses brown and 3 horses white. 7 brown horses went to the barn, so 3 white horses were still in the yard!

Lesson 41:   Culminating task—choose tools strategically to model and represent a stick of 10 cubes broken into two parts.   163

© 2018 Great Minds®. eureka-math.org

Name _____   Date _____

Complete a number bond and number sentence for each problem.

Color 6 blocks blue. Color the rest red. All of the blue blocks fell off the table. How many blocks are still on the table?

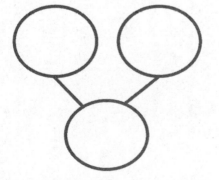

_____ - _____ = _____

Color some blocks orange and the rest yellow to make 10. All of the yellow blocks fell off the table. How many blocks are left?

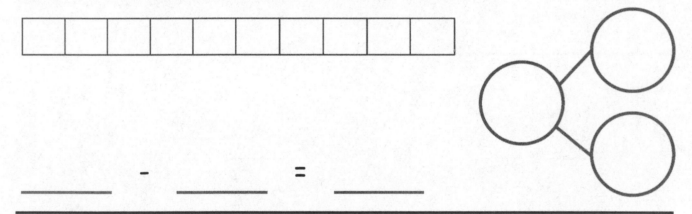

_____ - _____ = _____

Draw 5 dogs and some cats the 5-group way.

_____ + _____ = _____

Lesson 41:   Culminating task—choose tools strategically to model and represent a
              stick of 10 cubes broken into two parts.

© 2018 Great Minds®. eureka-math.org

165

There were 10 horses in the yard. Some were brown, and some were white.
Draw the horses the 5-group way. The brown ones went back into the
barn. How many horses were still in the yard? Draw a number bond, and
write a subtraction sentence.

Solve.

$1 + 1 =$ _____ $1 + 2 =$ _____

$2 + 1 =$ _____   _____ $= 2 + 2$

$3 + 1 =$ _____ $1 + 4 =$ _____

$4 + 1 =$ _____   _____ $= 3 + 2$

$5 + 1 =$ _____ $2 + 3 =$ _____

Lesson 41:   Culminating task—choose tools strategically to model and represent a
                                stick of 10 cubes broken into two parts.

© 2018 Great Minds®. eureka-math.org

Grade K
Module 5

Circle 10. Count the number of times you circled 10 ones. Tell a friend or an adult how many times you circled 10 ones.

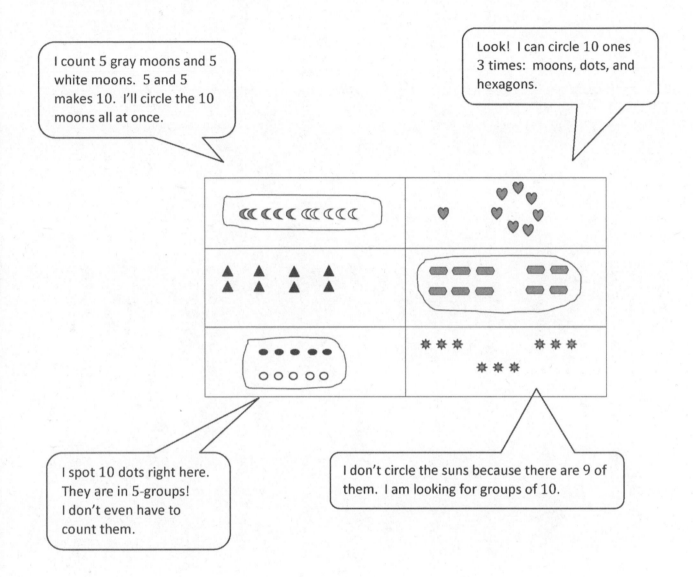

I count 5 gray moons and 5 white moons. 5 and 5 makes 10. I'll circle the 10 moons all at once.

Look! I can circle 10 ones 3 times: moons, dots, and hexagons.

I spot 10 dots right here. They are in 5-groups! I don't even have to count them.

I don't circle the suns because there are 9 of them. I am looking for groups of 10.

Name _____   Date _____

Circle 10.

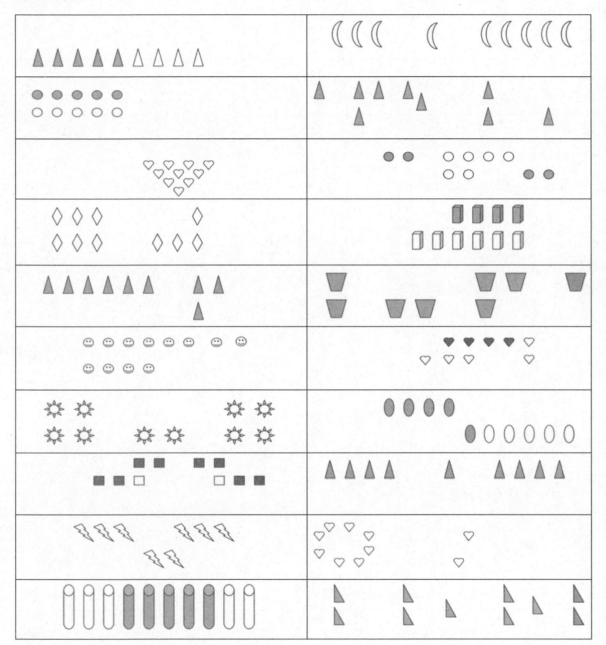

Count the number of times you circled 10 ones. Tell a friend or an adult how many times you circled 10 ones.

Lesson 1:   Count straws into piles of ten; count the piles as 10 ones.

171

© 2018 Great Minds®. eureka-math.org

Draw more to show the number.

10 ones and 3 ones

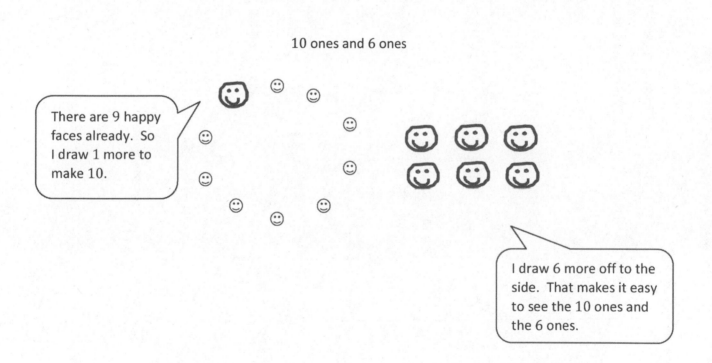

It's easy to see 10 dots right here. They are in 5-groups! So I just draw 3 more.

10 ones and 6 ones

There are 9 happy faces already. So I draw 1 more to make 10.

I draw 6 more off to the side. That makes it easy to see the 10 ones and the 6 ones.

Lesson 2: Count 10 objects within counts of 10 to 20 objects, and describe as 10 ones and __ ones.

173

© 2018 Great Minds®. eureka-math.org

Name _____   Date _____

△△△△△
△△△△△   △△ △

10 ones and 3 ones

Draw more to show the number.

◯ ◯ ◯ ◯ ◯

◯ ◯ ◯ ◯ ◯

10 ones and 2 ones

♡ ♡ ♡ ♡ ♡ ♡ ♡ ♡ ♡ ♡

♡ ♡ ♡ ♡

10 ones and 5 ones

(((((

(((((

10 ones and 7 ones

10 ones and 4 ones

Lesson 2:   Count 10 objects within counts of 10 to 20 objects, and describe as 10 ones and __ ones.

175

© 2018 Great Minds®. eureka-math.org

Circle 10 things. Tell how many there are in two parts, 10 ones and some more ones.

It's easy to find the 10 ones when they are in 5-groups.

I circle 10 ones and count the rest. Here are 4 more ones.

I have 10 ones and 4 ones.

It's a little tricky to find the 10 ones here. I make a line so that I remember where I start counting and then keep going around until I get to 10.

I have 10 ones and 3 ones.

EUREKA MATH   **Lesson 3:**   Count and circle 10 objects within images of 10 to 20 objects, and   **177**
                              describe as 10 ones and __ ones.

© 2018 Great Minds®. eureka-math.org

Name _____   Date _____

I have 10 ones and 3 ones.

Circle 10 things. Tell how many there are in two parts, 10 ones and some more ones.

I have 10 ones and ____ ones.

I have 10 ones and ____ ones.

I have ____ ones and ____ ones.

I have ____ ones and ____ ones.

EUREKA MATH®

Lesson 3:   Count and circle 10 objects within images of 10 to 20 objects, and describe as 10 ones and __ ones.

179

© 2018 Great Minds®. eureka-math.org

Draw a line to match each picture with the numbers the Say Ten way.

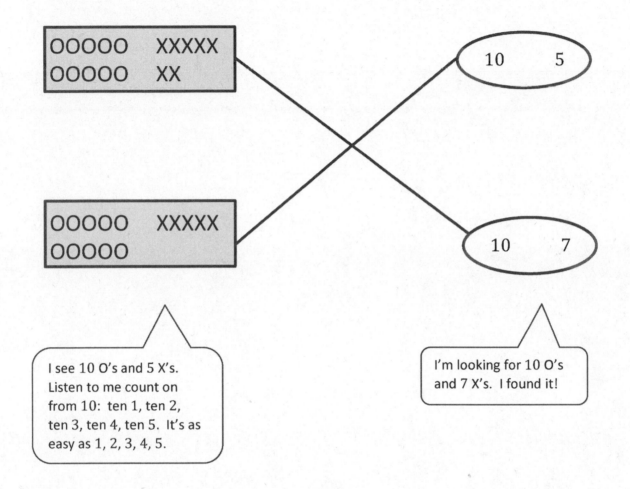

I see 10 O's and 5 X's. Listen to me count on from 10: ten 1, ten 2, ten 3, ten 4, ten 5. It's as easy as 1, 2, 3, 4, 5.

I'm looking for 10 O's and 7 X's. I found it!

Name _____   Date _____

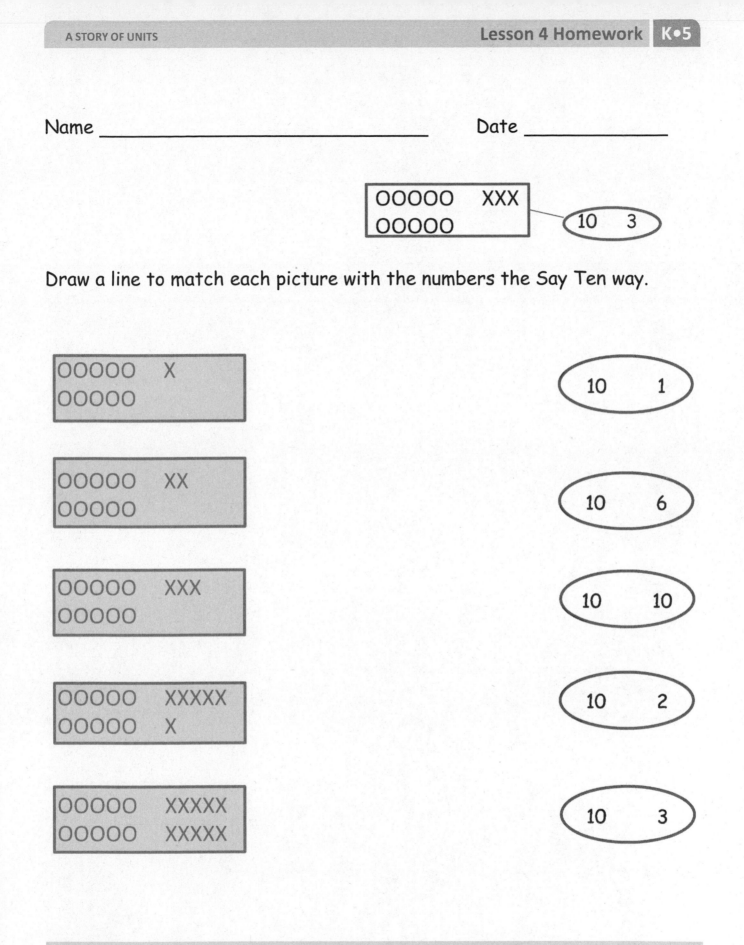

Draw a line to match each picture with the numbers the Say Ten way.

© 2018 Great Minds®. eureka-math.org

Write the numbers that go before and after, counting the Say Ten way.

> Putting "and" in the middle helps me think of the number in two parts.

> I can count the Say Ten way: ten 1, ten 2, ten 3, ten 4, ten 5, ten 6, ten 7, ten 8, ten 9, 2 ten. Another way to say 2 ten is 10 and 10.

BEFORE	NUMBER	AFTER
10 and 2	10 and 3	10 and 4
10 and 6	10 and 7	10 and 8
10 and 7	10 and 8	10 and 9

> I just count the Say Ten way and listen for the numbers before and after. Then I know what to write!

Lesson 5:   Count straws the Say Ten way to 20; make a pile for each ten.

185

EUREKA MATH

© 2018 Great Minds®. eureka-math.org

Name _____   Date _____

Write the numbers that go before and after, counting the Say Ten way.

BEFORE	NUMBER	AFTER
10 and 3	10 and 4	10 and 5
and	10 and 2	and
and	10 and 5	and
and	10 and 6	and
and	10 and 1	and
and	10 and 9	and

Lesson 5:   Count straws the Say Ten way to 20; make a pile for each ten.

187

© 2018 Great Minds®. eureka-math.org

Write and draw the number. Use your Hide Zero cards to help you.

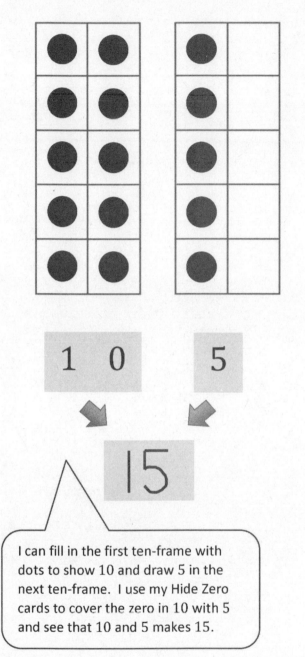

1 0   5

15

I can fill in the first ten-frame with dots to show 10 and draw 5 in the next ten-frame. I use my Hide Zero cards to cover the zero in 10 with 5 and see that 10 and 5 makes 15.

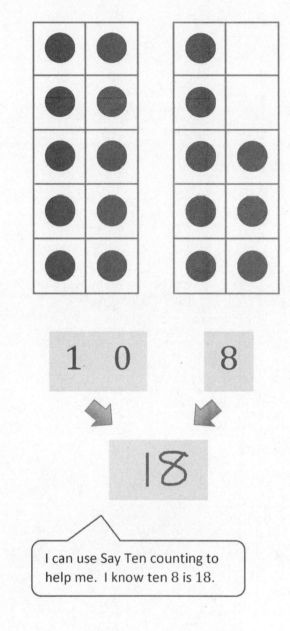

1 0   8

18

I can use Say Ten counting to help me. I know ten 8 is 18.

Lesson 6:   Model with objects and represent numbers 10 to 20 with place value or Hide Zero cards.

189

© 2018 Great Minds®. eureka-math.org

Name _____   Date _____

Write and draw the number. Use your Hide Zero cards to help you.

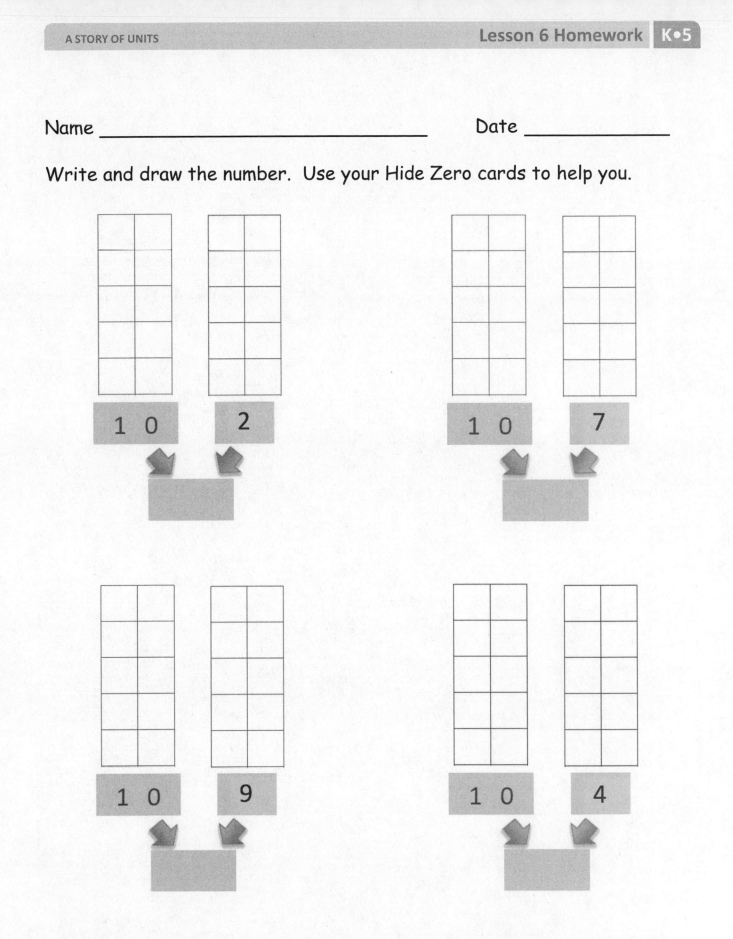

Lesson 6:   Model with objects and represent numbers 10 to 20 with place value or
Hide Zero cards.

191

EUREKA MATH

© 2018 Great Minds®. eureka-math.org

Note: Match to corresponding 5-group side and copy double-sided on card stock.

large Hide Zero cards (numeral side)

Lesson 6: Model with objects and represent numbers 10 to 20 with place value or Hide Zero cards.

193

© 2018 Great Minds®. eureka-math.org

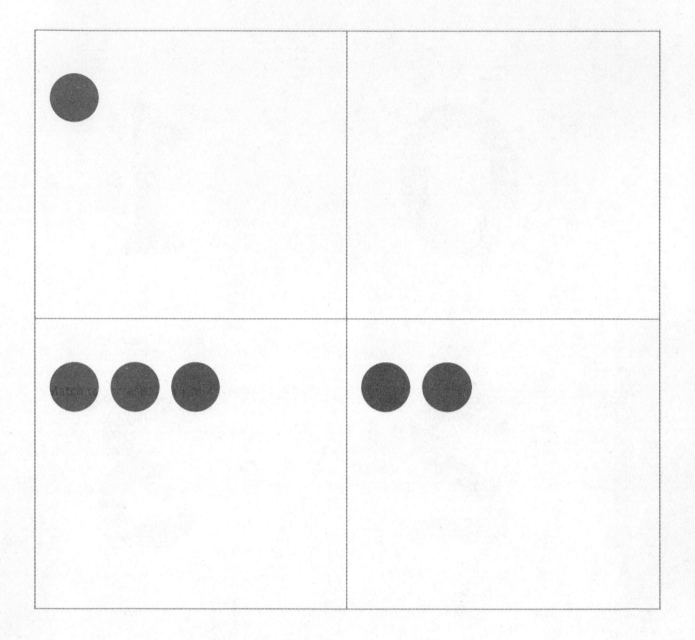

Note: Match to corresponding numeral side and copy double-sided on card stock.

large Hide Zero cards (5-group side)

Lesson 6:   Model with objects and represent numbers 10 to 20 with place value or
                        Hide Zero cards.

© 2018 Great Minds®. eureka-math.org

Note: Match to corresponding 5-group side and copy double-sided on card stock.

large Hide Zero cards (numeral side)

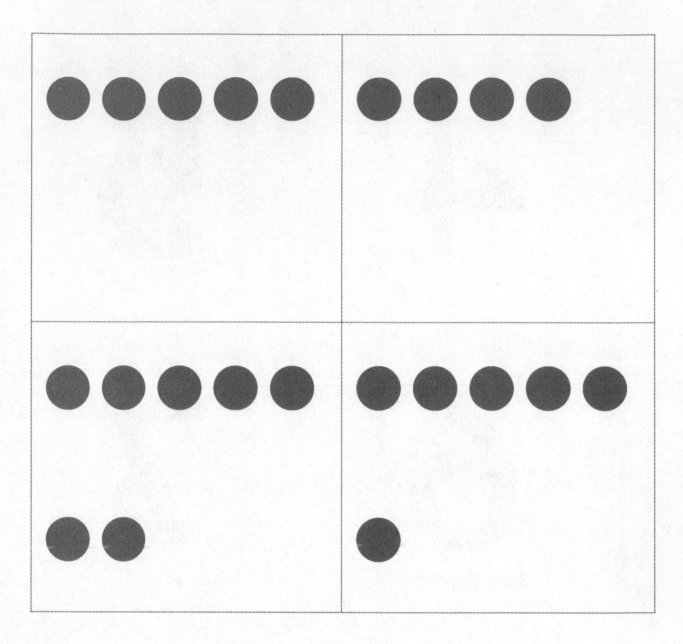

Note: Match to corresponding numeral side and copy double-sided on card stock.

large Hide Zero cards (5-group side)

Lesson 6:   Model with objects and represent numbers 10 to 20 with place value or
                            Hide Zero cards.

© 2018 Great Minds®. eureka-math.org

Note: Match to corresponding 5-group side and copy double-sided on card stock.

large Hide Zero cards (numeral side)

Lesson 6:   Model with objects and represent numbers 10 to 20 with place value or
              Hide Zero cards.

197

© 2018 Great Minds®. eureka-math.org

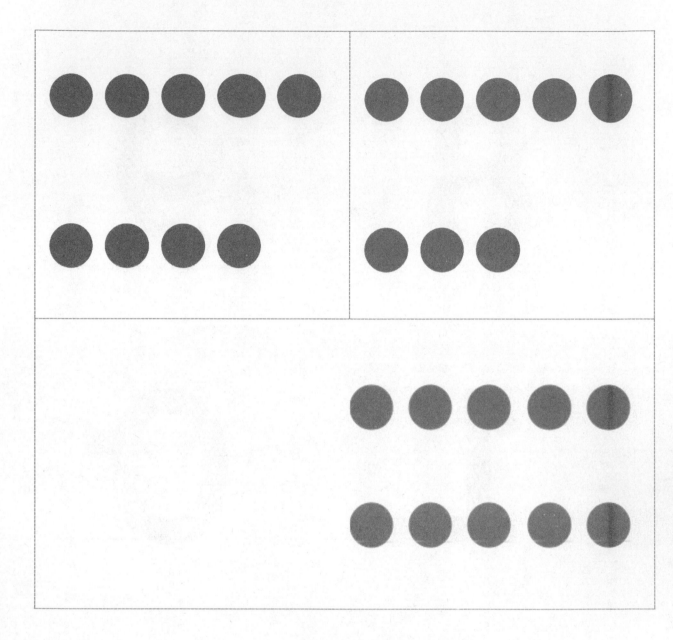

Note: Match to corresponding numeral side and copy double-sided on card stock.

large Hide Zero cards (5-group side)

Lesson 6: Model with objects and represent numbers 10 to 20 with place value or Hide Zero cards.

© 2018 Great Minds®. eureka-math.org

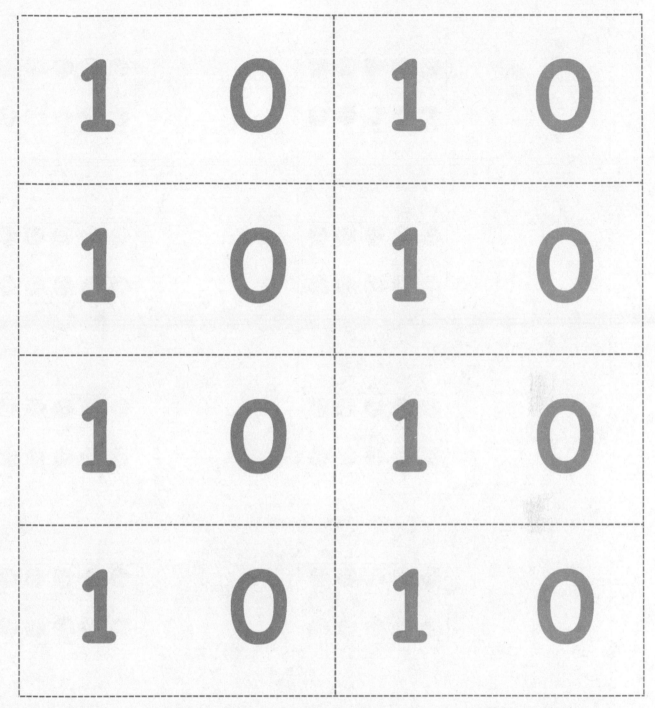

Note: Copy double-sided with the Hide Zero 10 card (5-group side) on card stock. Each student needs one, double-sided Hide Zero 10 card. This card is used with 5-group cards 1–9 (Lesson 1 Fluency Template 2), which combined, make the full set of Hide Zero cards.

_____

Hide Zero 10 card (numeral side)

EUREKA MATH®

Lesson 6:   Model with objects and represent numbers 10 to 20 with place value or Hide Zero cards.

© 2018 Great Minds®. eureka-math.org

199

Note: Copy double-sided with the Hide Zero 10 card (numeral side) on card stock. Each student needs one, double-sided Hide Zero 10 card. This card is used with 5-group cards 1–9 (Lesson 1 Fluency Template 2), which combined, make the full set of Hide Zero cards.

_____

Hide Zero 10 card (5-group side)

© 2018 Great Minds®. eureka-math.org

Look at the Hide Zero cards or the 5-group cards. Use your cards to show the number. Write the number as a number bond.

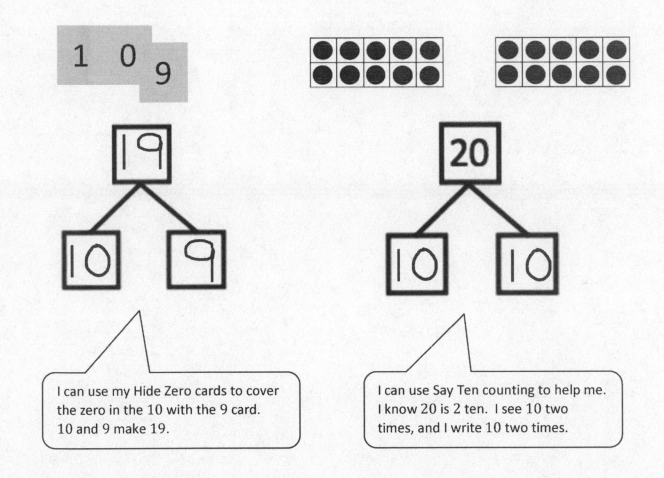

I can use my Hide Zero cards to cover the zero in the 10 with the 9 card. 10 and 9 make 19.

I can use Say Ten counting to help me. I know 20 is 2 ten. I see 10 two times, and I write 10 two times.

Name _____ Date _____

Look at the Hide Zero cards or the 5-group cards. Use your cards to show the number. Write the number as a number bond.

© 2018 Great Minds®. eureka-math.org

Use your materials to show each number as 10 ones and some more ones. Use your 5-groups way of drawing.

1 4

Ten six

I know 14 is 10 and 4. I can use pennies to show 14. I put down 10 pennies the 5-group way. That's easy. 5 and 5 makes 10. Then I just put 4 more. I can draw a picture of my pennies.

Ten six is the Say Ten counting way to say 16. This time I can use cereal to show 16. I can draw 16 circles to show how I arrange my o-shaped cereal. I see 10 ones and 6 more ones. I count them like this: ten 1, ten 2, ten 3, ten 4, ten 5, ten 6. I did it right!

Name _____   Date _____

Use your materials to show each number as 10 ones and some more ones.
Use your 5-groups way of drawing.

1   5

1   3

Ten seven

Ten one

© 2018 Great Minds®. eureka-math.org

1   2

1   6

2 ten

Ten four

Lesson 8:   Model teen numbers with materials from abstract to concrete.

© 2018 Great Minds®. eureka-math.org

For each number, make a drawing that shows that many objects. Circle 10 ones.

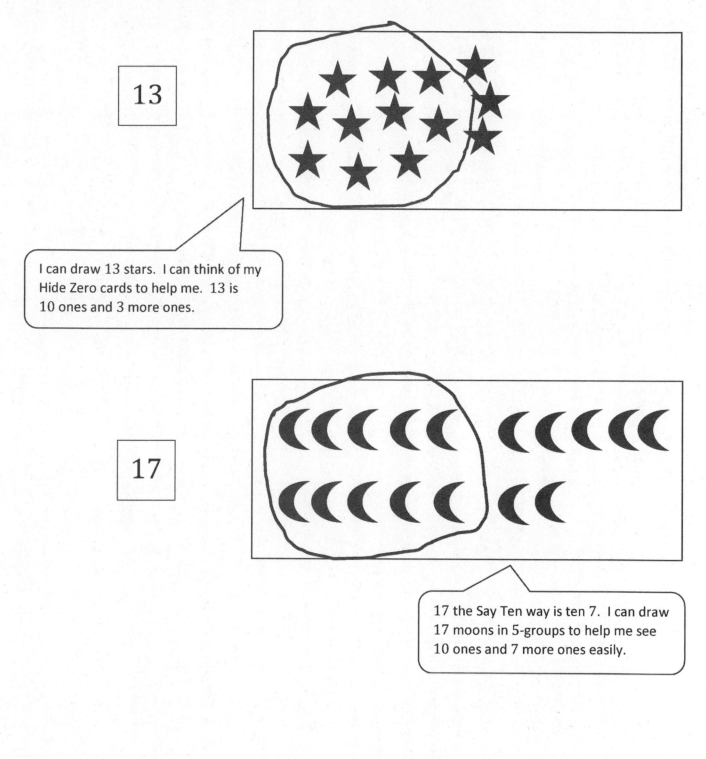

13

I can draw 13 stars. I can think of my Hide Zero cards to help me. 13 is 10 ones and 3 more ones.

17

17 the Say Ten way is ten 7. I can draw 17 moons in 5-groups to help me see 10 ones and 7 more ones easily.

EUREKA MATH

© 2018 Great Minds®. eureka-math.org

Name _____   Date _____

For each number, make a drawing that shows that many objects.
Circle 10 ones.

11

16

20

© 2018 Great Minds®. eureka-math.org

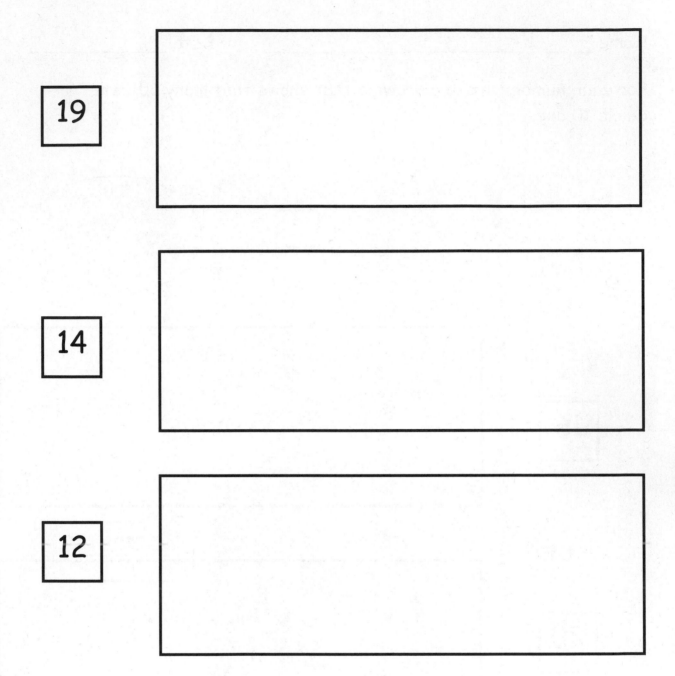

19

14

12

Lesson 9:   Draw teen numbers from abstract to pictorial.

© 2018 Great Minds®. eureka-math.org

Color the number of fingernails and beads to match the number bond. Show by coloring 10 ones above and extra ones below. Fill in the number bonds.

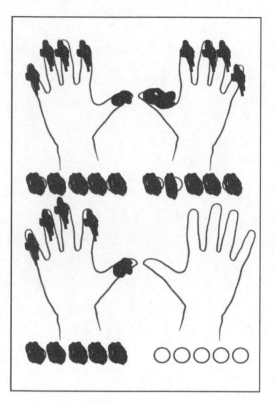

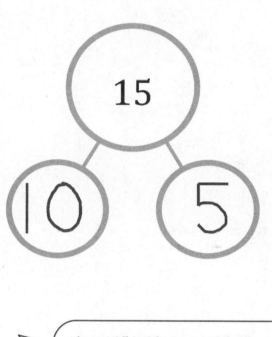

I know 15 is 10 ones and 5 ones. I can color 10 fingernails and beads on top. I can color 5 more fingernails and beads below. I fill in the number bond with 10 and 5 to match my drawing.

Name _____   Date _____

Color the number of fingernails and beads to match the number bond. Show by coloring 10 ones above and extra ones below. Fill in the number bonds.

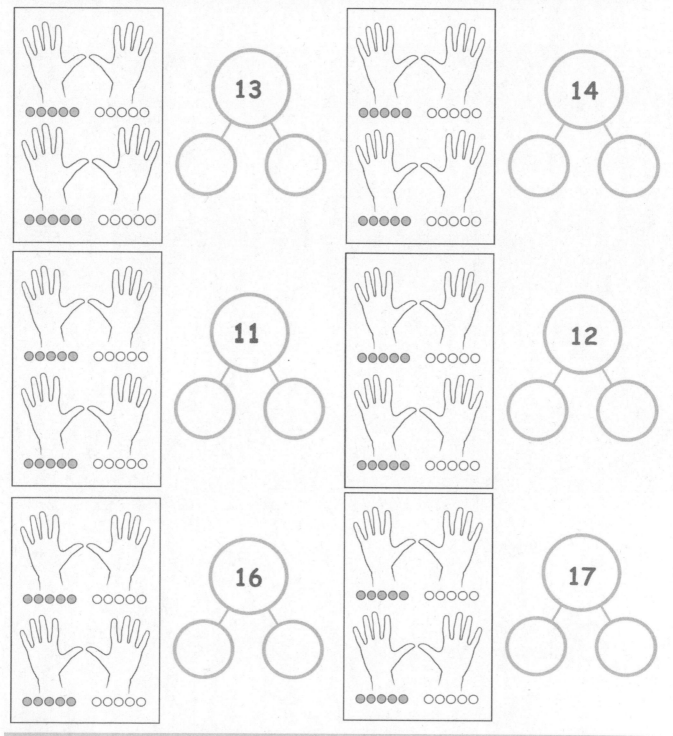

EUREKA MATH®

© 2018 Great Minds®. eureka-math.org

Write the missing numbers. Then, count and draw X's and O's to complete the pattern.

10	11	12	13	14	15	16	17	18	19	20

To find the missing number, I use the pattern of 1 larger. It goes like this:

10. 1 more is 11.

11. 1 more is 12.

I draw 10 O's and 2 X's. Ten 2 is the same as 12.

I can think of my Hide Zero cards and Say Ten counting, too. I know 19 is 10 ones and 9 more ones. I can draw 10 O's and 9 X's.

Lesson 11: Show, count, and write numbers 11 to 20 in tower configurations increasing by 1—a pattern of 1 larger.

217

© 2018 Great Minds®. eureka-math.org

Name _____ Date _____

Write the missing numbers. Then, count and draw X's and O's to complete the pattern.

10		12		14		16	17	18		20

Lesson 11: Show, count, and write numbers 11 to 20 in tower configurations increasing by 1—a pattern of *1 larger*.

219

© 2018 Great Minds®. eureka-math.org

Write the missing numbers. Then, draw X's and O's to complete the pattern.

20	19	18	17	16	15	14	13	12	11	10

I count the O's and X's. There are 10 O's and 10 X's. That's 2 ten. 2 ten is the same as 20.

I know I'm on the right track because I hear the pattern of 1 smaller. It goes like this:

14. 1 less is 13.

13. 1 less is 12.

12. 1 less is 11.

Name _____   Date _____

Write the missing numbers. Then, draw X's and O's to complete the pattern.

X										
X	X									
X	X									
X	X		X							
X	X		X							
X	X		X	X						
X	X		X	X						
X	X		X	X						
X	X		X	X				X	
X	X		X	X				O	
O	O		O	O				O	
O	O		O	O				O	
O	O		O	O				O	
O	O		O	O				O	
O	O		O	O				O	
O	O		O	O				O	
O	O		O	O				O	
O	O		O	O				O	
O	0		O	O				O	
20		**18**		**16**		**14**	**13**	**12**		**10**

Count the objects. Draw dots to show the same number on the double 10-frames.

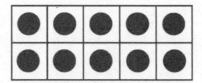

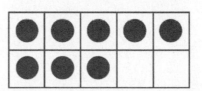

I can count the stars. I point to each one as I count. There are 18 stars.

I know 18 the Say Ten way is ten 8. I can fill in the top frame with ten ones and draw 8 more ones in the bottom ten-frame. I can draw 8 ones easily. I know 8 is five and three.

Lesson 13:   Show, count, and write to answer *how many* questions in linear and array configurations.

© 2018 Great Minds®. eureka-math.org

225

Name _____ Date _____

Count the objects. Draw dots to show the same number on the double 10-frames.

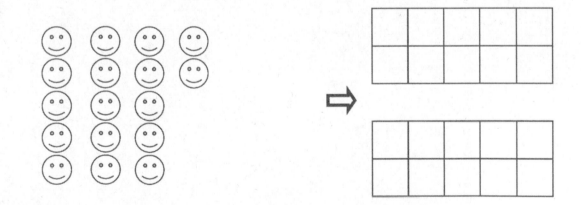

Lesson 13:   Show, count, and write to answer *how many* questions in linear and
array configurations.

227

© 2018 Great Minds®. eureka-math.org

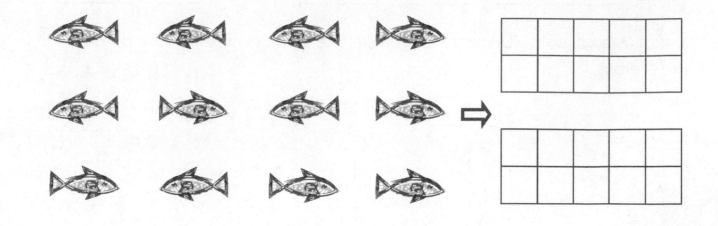

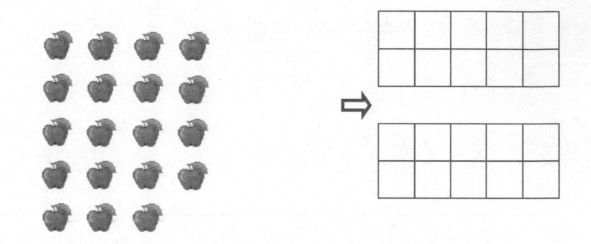

Lesson 13:   Show, count, and write to answer *how many* questions in linear and array configurations.

© 2018 Great Minds®. eureka-math.org

EUREKA
MATH

Count the objects. Write the number in the box next to the picture.

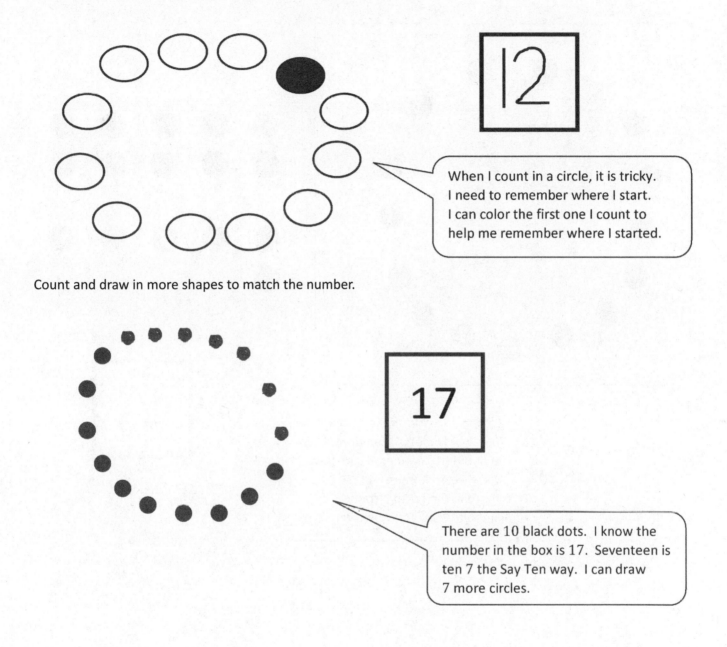

12

When I count in a circle, it is tricky.
I need to remember where I start.
I can color the first one I count to
help me remember where I started.

Count and draw in more shapes to match the number.

17

There are 10 black dots. I know the
number in the box is 17. Seventeen is
ten 7 the Say Ten way. I can draw
7 more circles.

Count the dots. Draw each dot in the 10-frame. Write the number in the box below the 10-frames.

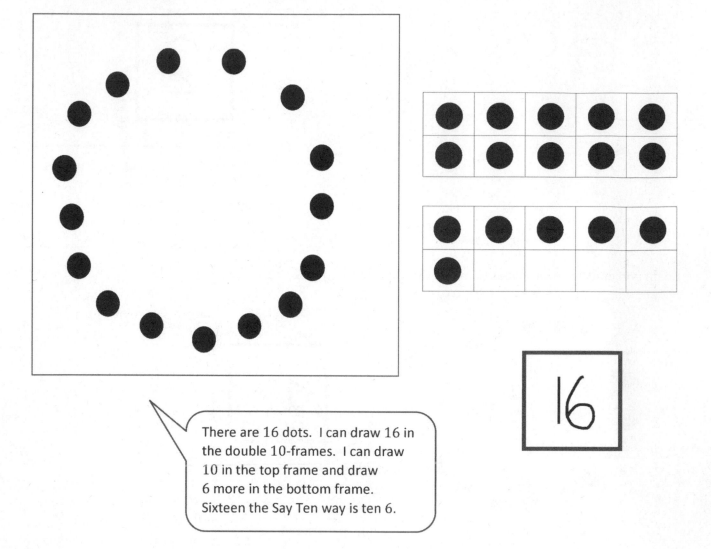

There are 16 dots. I can draw 16 in the double 10-frames. I can draw 10 in the top frame and draw 6 more in the bottom frame. Sixteen the Say Ten way is ten 6.

16

Lesson 14:   Show, count, and write to answer *how many* questions with up to 20 objects in circular configurations.

© 2018 Great Minds®. eureka-math.org

Name _____ Date _____

Count the objects in each group. Write the number in the boxes below the pictures.

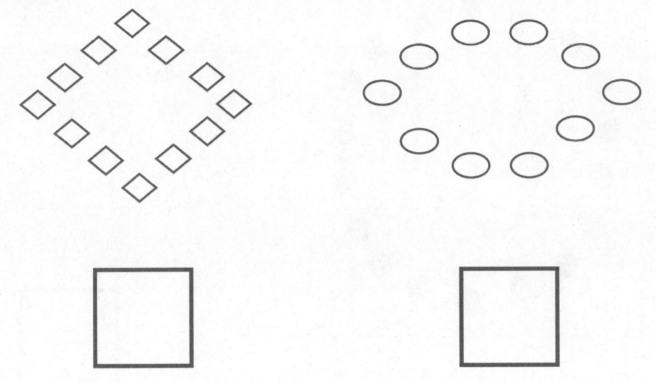

Count and draw in more shapes to match the number.

19

EUREKA MATH®

Lesson 14:   Show, count, and write to answer *how many* questions with up to 20 objects in circular configurations.

© 2018 Great Minds®. eureka-math.org

231

Count the dots. Draw each dot in the 10-frame. Write the number in the box below the 10-frames.

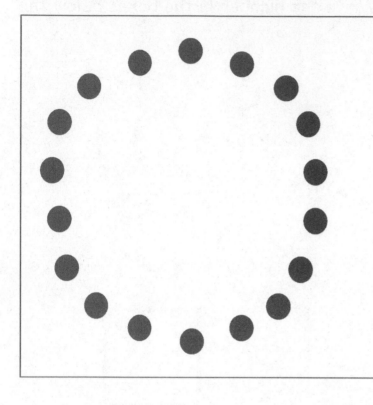

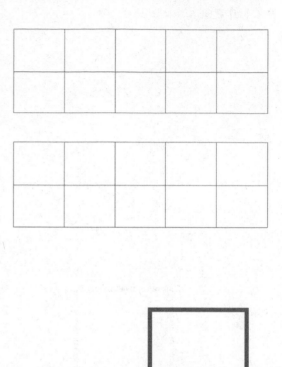

Write a teen number in the box below. Draw a picture to match your number.

Lesson 14:   Show, count, and write to answer *how many* questions with up to
                              20 objects in circular configurations.

© 2018 Great Minds®. eureka-math.org

EUREKA
MATH®

Count the Say Ten way. Write the missing numbers.

(ten-frames)	60	__6__ tens
(ten-frames)	70	7 tens
(ten-frames)	80	__8__ tens
(ten-frames)	70	__7__ tens
(ten-frames)	60	6 tens

> I can count by tens and the Say Ten way! I count the ten-frames first. There are 6 ten-frames, so that is 6 tens. 6 tens is the same as 60.

EUREKA MATH®

© 2018 Great Minds®. eureka-math.org

Name _____ Date _____

Count down by 10, and write the number on top of each stair.

100

40

10

© 2018 Great Minds®. eureka-math.org

Count down the Say Ten way. Write the missing numbers.

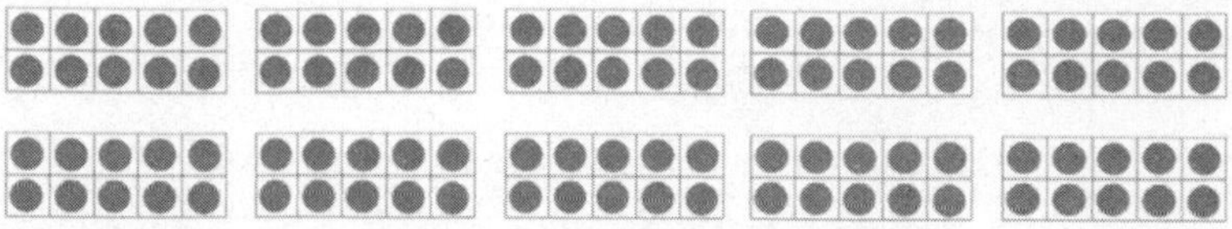	100	
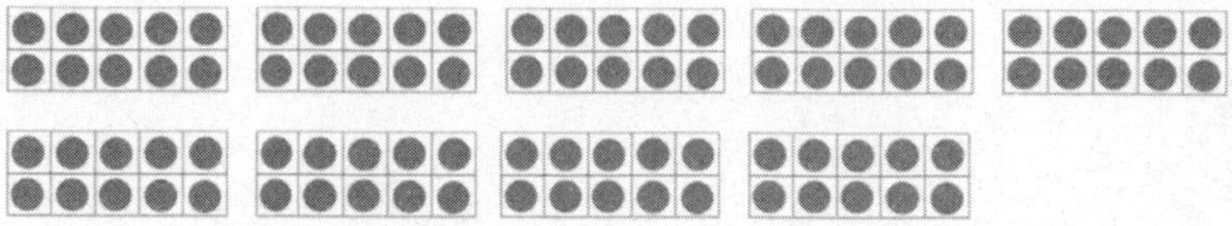		9 tens
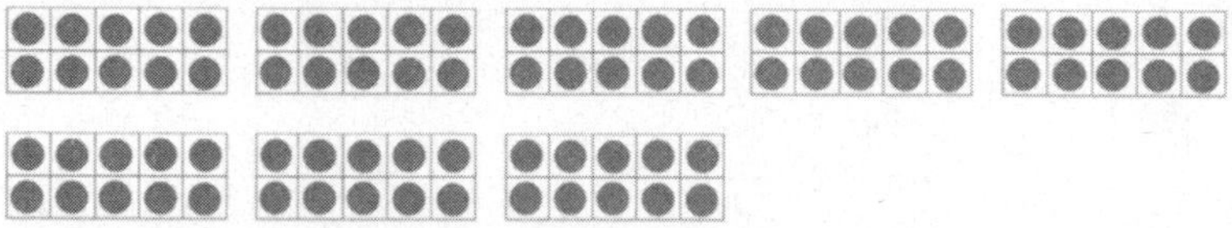	80	_____ tens
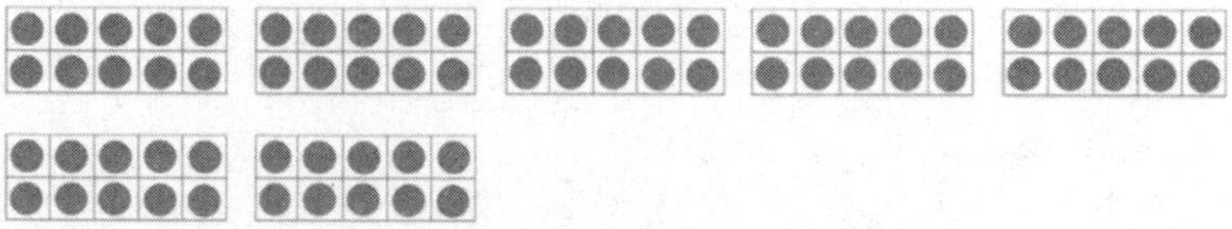	70	_____ tens
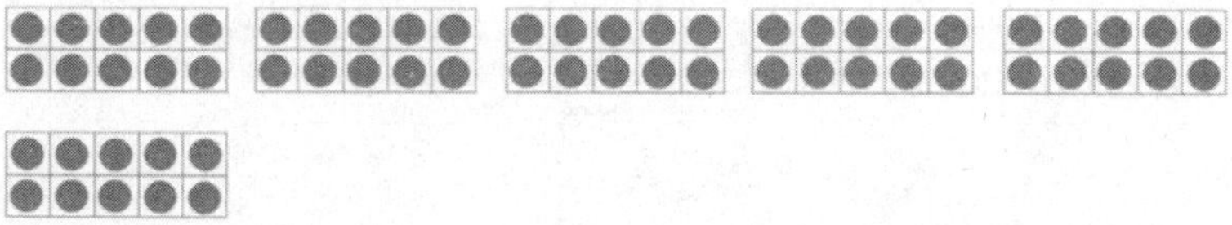		6 tens
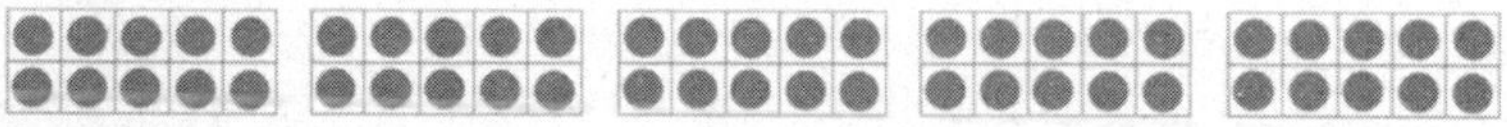		_____ tens
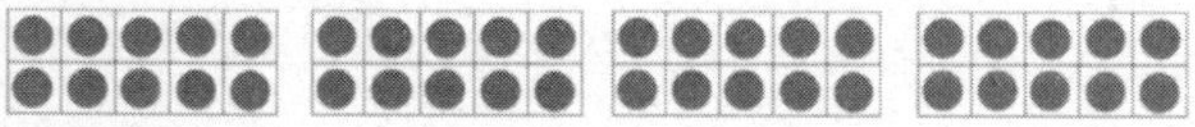	40	4 tens
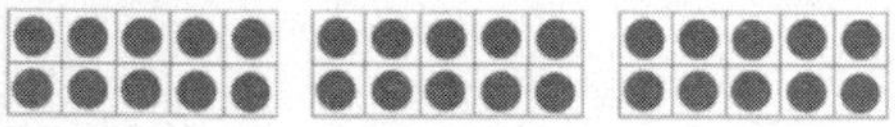		_____ tens
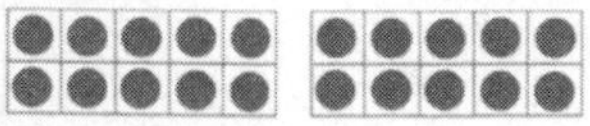		_____ tens
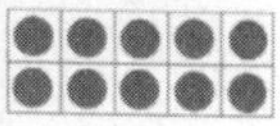		_____ ten

© 2018 Great Minds®. eureka-math.org

EUREKA MATH

Help the rabbit get his carrot. Count by 1's.

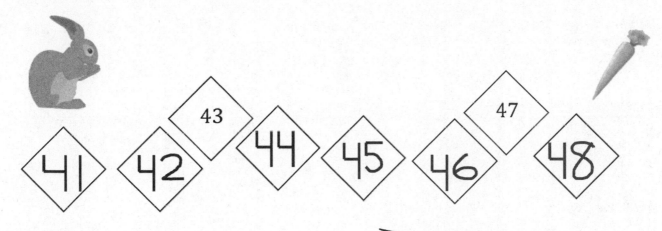

Count up by 1's and then down by 1's.

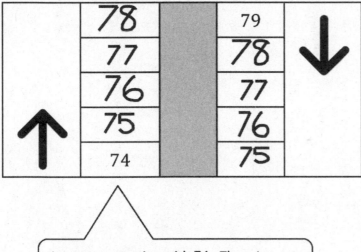

I help the rabbit get to the carrot by counting by 1's. I count backward from 43 to fill in 42 and 41. Then, I count forward from 43 to fill in the rest of the numbers.

I count up starting with 74. Then, I count down in the next column from 79.

Name _____   Date _____

Help the rabbit get his carrot. Count by 1s.

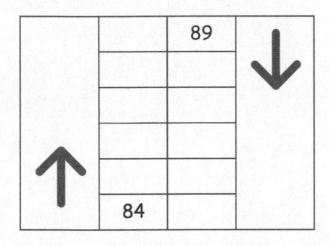

71   75

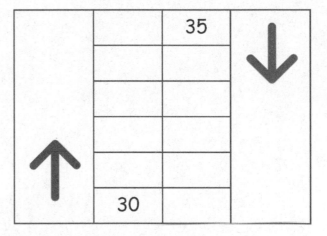

12   14   16

Count up by 1s, then down by 1s.

		89	↓
↑			
	84		

		35	↓
↑			
	30		

Help the boy mail his letter. Count up by 1s. When you get to the top, count down by 1s.

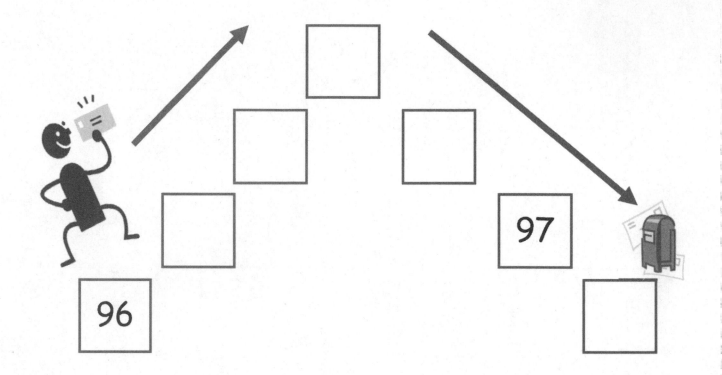

96

97

Lesson 16:   Count within tens by ones.

© 2018 Great Minds®. eureka-math.org

Draw more to show the number.

42 is the same as 4 tens 2. The first ten-frame is full, so I don't need to draw more dots. I make dots in each ten-frame until 4 ten-frames are full. Then, I add two more dots to make 42.

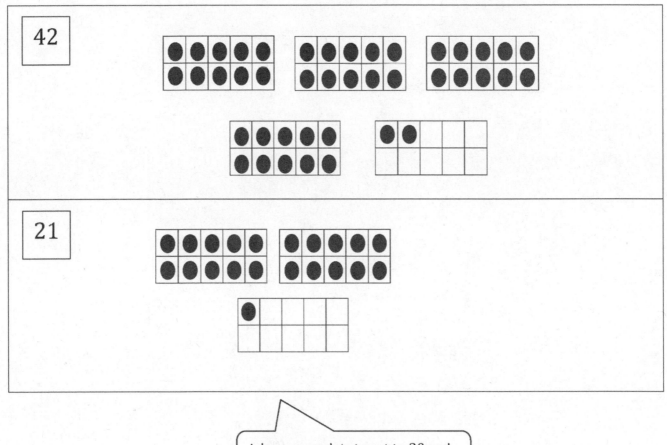

I draw more dots to get to 20 and then add 1 more to make 21 dots!

Name _____   Date _____

Draw more to show the number.

Example:

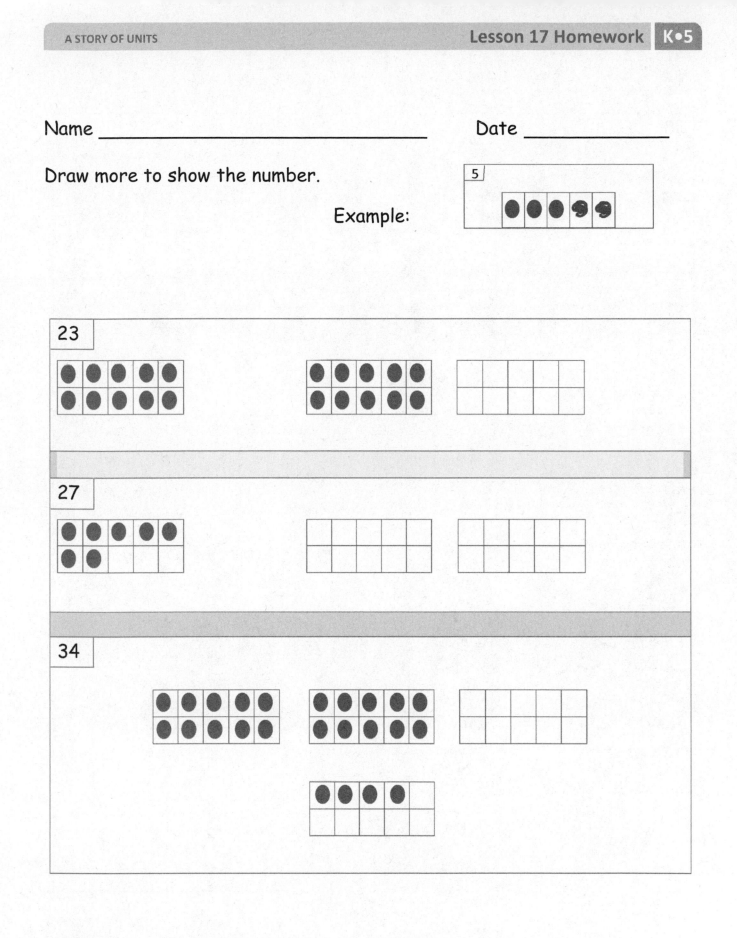

Lesson 17:   Count across tens when counting by ones through 40.

243

© 2018 Great Minds®. eureka-math.org

38

40

Lesson 17:   Count across tens when counting by ones through 40.

© 2018 Great Minds®. eureka-math.org

Use your Rekenrek, hiding paper (a blank sheet of paper), and crayons to complete each step listed below. Read and complete the problems with the help of an adult.

Hide to show just 40 on your Rekenrek dot paper. Touch and count the circles until you say 24. Color 24 (the 24th circle) green.

- Touch and count each circle from 24 to 32.
- Color 32 (the 32nd circle) with a red crayon.

© 2018 Great Minds®. eureka-math.org

Directions for Rekenrek Homework

Use your Rekenrek (attached), hiding paper (an extra paper to hide some of the dots), and crayons to complete each step listed below. Read and complete the problems with the help of an adult.

Hide to show just 40 on your Rekenrek dot paper. Touch and count the circles until you say 28. Color 28 green.

- Touch and count each circle from 28 to 34.
- Color 34 (the 34th circle) with a red crayon

_____

Hide to show just 60 on your Rekenrek dot paper. Touch and count the circles until you say 45. Color 45 yellow.

- Touch and count each circle from 45 to 52.
- Color 52 with a blue crayon.

_____

Hide to show just 90 on your Rekenrek dot paper. Touch and count the circles until you say 83. Color 83 purple.

- Touch and count down from 83 to 77.
- Color 77 with a red crayon.

_____

Show 100.

- Touch and count, starting at 1.
- Say the last number in each row loudly. Color the circle black.

EUREKA MATH

Lesson 18:   Count across tens by ones to 100 with and without objects.

247

© 2018 Great Minds®. eureka-math.org

Name _____   Date _____

Rekenrek

© 2018 Great Minds®. eureka-math.org

Write the number you see. Now, draw one more. Then write the new number.

> I count 30 smiley faces.
> I draw 1 more smiley
> face, and now there are
> 31 smiley faces.

> I see 4 full ten-frames and 7 dots.
> That is 4 tens 7. I add a dot, and now
> there are 4 tens 8, which is 48.

Lesson 19:   Explore numbers on the Rekenrek. (Optional)   249

EUREKA MATH

© 2018 Great Minds®. eureka-math.org

Name _____   Date _____

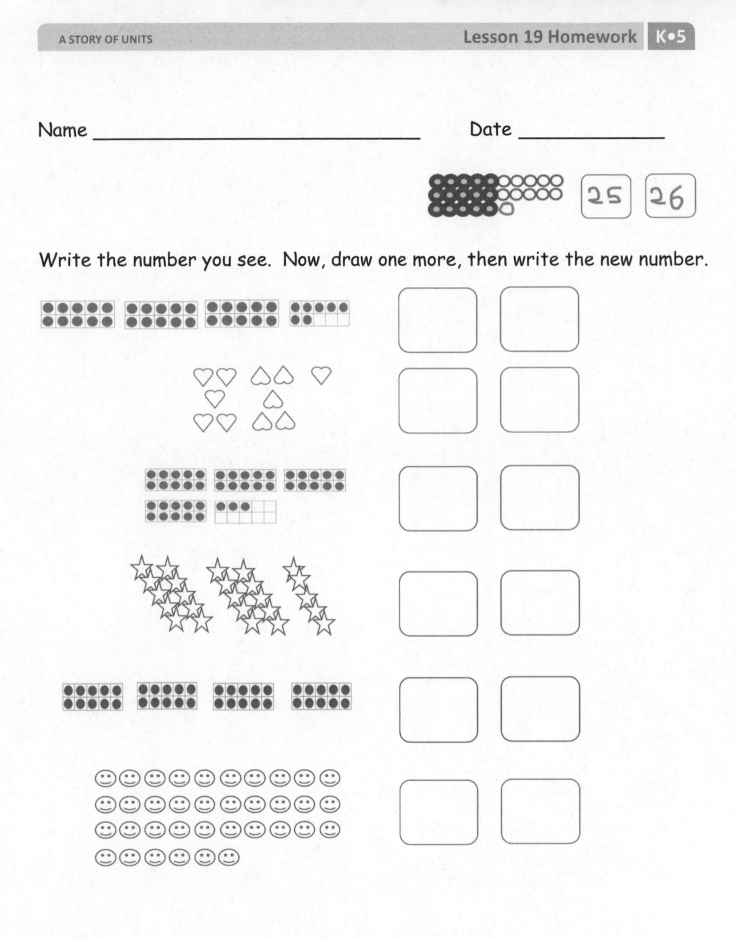

Write the number you see. Now, draw one more, then write the new number.

Lesson 19:   Explore numbers on the Rekenrek. (Optional)

251

EUREKA
MATH

© 2018 Great Minds®. eureka-math.org

Draw stars to show the number as a number bond of 10 ones and some ones. Show each example as two addition sentences of 10 ones and some ones.

```
* * * * *          * * * * *
* * * * *              *
```

16

I need to show 16 stars!
There are 10 stars, so I draw 6 more to show my two parts.

$$10 + 6 = 16$$

$$16 = 10 + 6$$

I can make two addition sentences! I show my two parts in the first addition sentence. For the second number sentence, I show the whole first and then the parts.

Name _____   Date _____

Draw stars to show the number as a number bond of 10 ones and some ones. Show each example as two addition sentences of 10 ones and some ones.

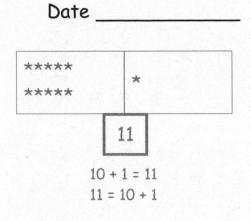

10 + 1 = 11
11 = 10 + 1

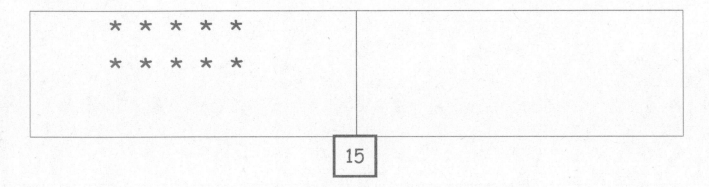

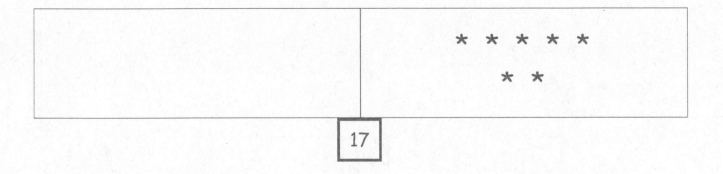

Lesson 20:   Represent teen number compositions and decompositions as addition sentences.

© 2018 Great Minds®. eureka-math.org

EUREKA MATH®

255

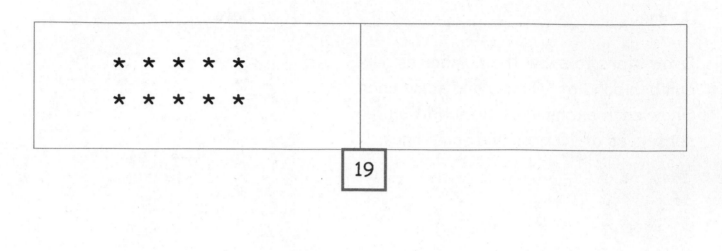

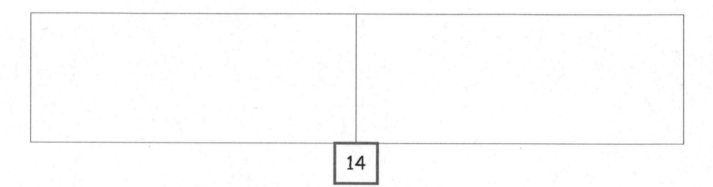

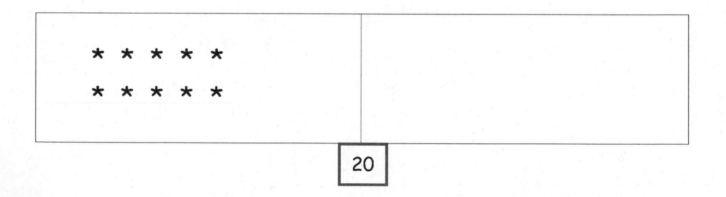

Lesson 20: Represent teen number compositions and decompositions as addition sentences.

© 2018 Great Minds®. eureka-math.org

Complete the number bond and number sentence. Draw the cubes of the missing part.

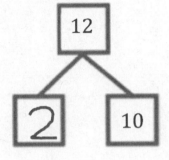

$$12 = \underline{2} + 10$$

> Hiding inside of 12 are 10 ones and 2 ones. I write a 2 to finish the number bond and the number sentence. There are 10 cubes already there, so I draw 2 more cubes to make ten 2, or 12 cubes.

Lesson 21:   Represent teen number decompositions as 10 ones and some ones, and find a hidden part.

257

© 2018 Great Minds®. eureka-math.org

Name _____ Date _____

Complete the number bonds and number sentences. Draw the cubes of the missing part.

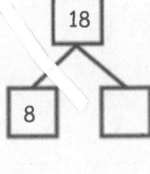

15 = _____ + 10

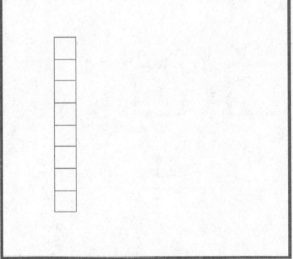

18

8

_____ + 8 = 18

Lesson 21: Represent teen number decompositions as 10 ones and some ones, and find a hidden part.

© 2018 Great Minds®. eureka-math.org

259

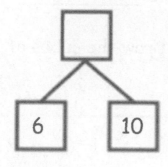

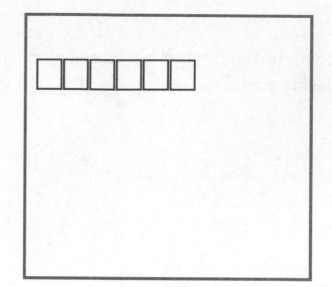

6 + _____ = 16

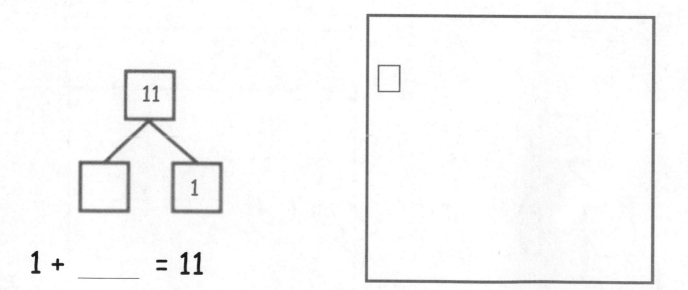

1 + _____ = 11

Lesson 21:   Represent teen number decompositions as 10 ones and some ones, and find a hidden part.

© 2018 Great Minds®. eureka-math.org

EUREKA MATH®

Fill in the number bond. Check the group with more.

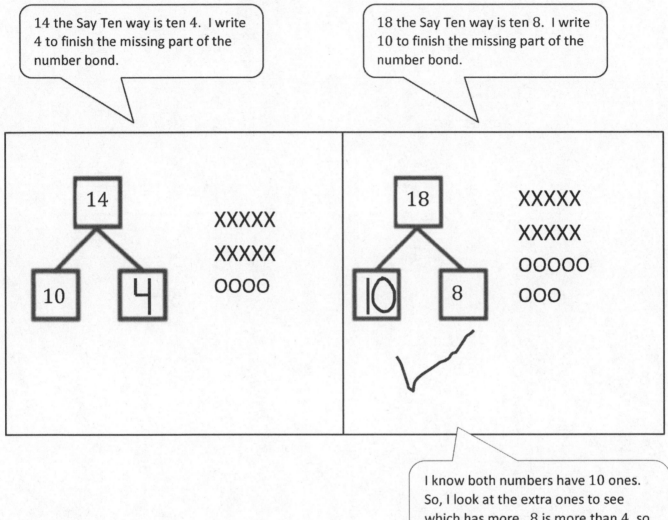

14 the Say Ten way is ten 4. I write 4 to finish the missing part of the number bond.

18 the Say Ten way is ten 8. I write 10 to finish the missing part of the number bond.

I know both numbers have 10 ones. So, I look at the extra ones to see which has more. 8 is more than 4, so that means ten 8 is more than ten 4.

EUREKA MATH®

© 2018 Great Minds®. eureka-math.org

Name _____   Date _____

Fill in the number bond.
Check the group with more.

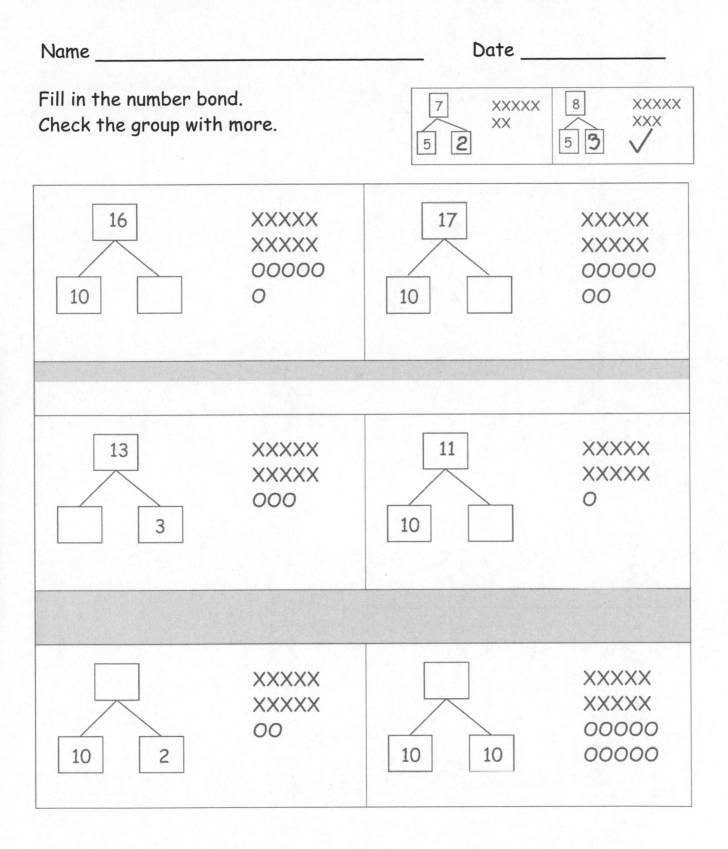

7
5 2   XXXXX
          XX

8
5 3   XXXXX
          XXX ✓

16 10 ☐	XXXXX XXXXX OOOOO O	17 10 ☐	XXXXX XXXXX OOOOO OO
13 ☐ 3	XXXXX XXXXX OOO	11 10 ☐	XXXXX XXXXX O
☐ 10 2	XXXXX XXXXX OO	☐ 10 10	XXXXX XXXXX OOOOO OOOOO

Lesson 22:   Decompose teen numbers as 10 ones and some ones; compare *some ones* to compare the teen numbers.

© 2018 Great Minds®. eureka-math.org

263

Bob bought 5 strawberry doughnuts and 10 chocolate doughnuts. Draw and show all of Bob's doughnuts.

Write an addition sentence to match your drawing.

$$\underline{5} + \underline{10} = \underline{15}$$

It's easy to see the doughnuts in two parts: strawberry and chocolate. 5 and 10 is the same as ten 5. That's 15.

I am great at making addition sentences! Let me tell you how my addition sentence matches my picture. The number 5 tells about the strawberry doughnuts. The number 10 tells about the chocolate. The number 15 tells how many doughnuts in all.

Name _____   Date _____

Bob bought 7 sprinkle donuts and 10 chocolate donuts. Draw and show all of Bob's donuts.

Write an addition sentence to match your drawing.

_____   _____   _____

Fill in the number bond to match your sentence.

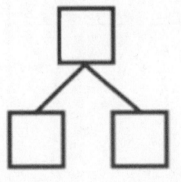

Fran has 17 baseball cards. Show Fran's baseball cards as 10 ones and some ones.

Write an addition sentence and a number bond that tell about the baseball cards.

_____ _____ _____

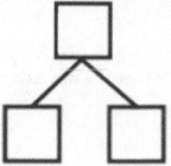

Lesson 23:   Reason about and represent situations, decomposing teen numbers into 10 ones and some ones and composing 10 ones and some ones into a teen number.
© 2018 Great Minds®. eureka-math.org

EUREKA MATH

Rabbit and Froggy's Matching Race

Directions: Play Rabbit and Froggy's Matching Race with a friend, relative, or parent to help your animal reach its food first! The first animal to reach the food wins.

- Put your teen numeral and dot cards face down in rows with teen numbers in one row and dot cards in another row.
- Flip to find 2 cards that match.
  Place cards back in the same place if they don't match.
  Continue until you find a match.

- Write a number bond to match. Hop 1 space if you get it right!

- Write a number sentence. Hop 1 space if you get it right!

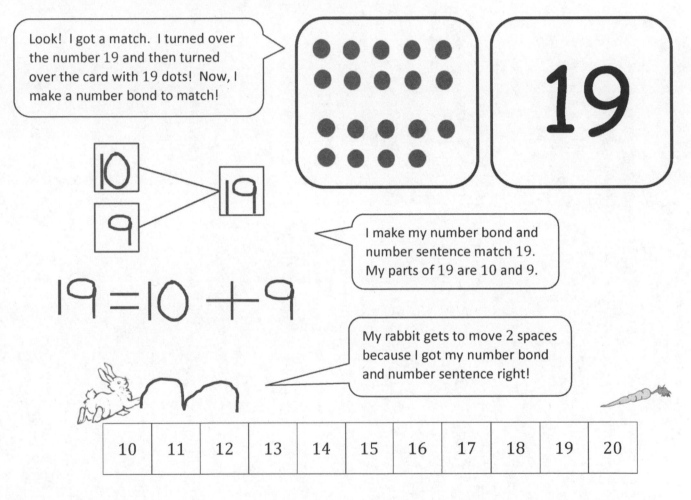

Look! I got a match. I turned over the number 19 and then turned over the card with 19 dots! Now, I make a number bond to match!

I make my number bond and number sentence match 19. My parts of 19 are 10 and 9.

$19 = 10 + 9$

My rabbit gets to move 2 spaces because I got my number bond and number sentence right!

| 10 | 11 | 12 | 13 | 14 | 15 | 16 | 17 | 18 | 19 | 20 |

Lesson 24:   Culminating Task—Represent teen number decompositions in various ways.

269

© 2018 Great Minds®. eureka-math.org

Rabbit and Froggy's Matching Race

Directions: Play Rabbit and Froggy's Matching Race with a friend, relative, or parent to help your animal reach its food first! The first animal to reach the food wins.

- Put your Teen number and Dot cards face down in rows with Teen numbers in one row and Dot cards in another row.

- Flip to find 2 cards that match.
  Place cards back in the same place if they don't match.
  Continue until you find a match.

- Write a number bond to match.   Hop 1 space if you get it right!

- Write a number sentence. $13 = 10 + 3$ Hop 1 space again if you get it right!

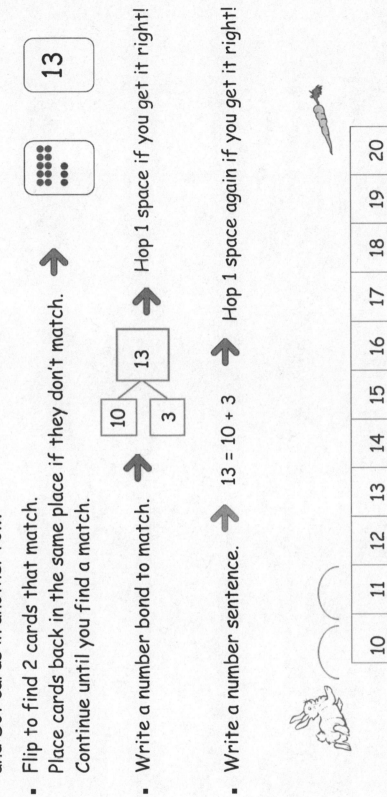

Grade K
Module 6

First, use your ruler to draw 2 lines to make a square. Second, color the corners red. Third, draw another square.

> I can follow directions! I use my ruler to draw 2 lines to finish the square. Then, I color the corners red.

> I can make a square! A square has 4 straight sides. I work hard to make sure the sides are all the same length.

Name _____   Date _____

Follow the directions.

First, use your ruler to draw a line finishing the triangle.

Second, color the triangle green.

Third, use your ruler to draw a bigger triangle next to the green triangle.

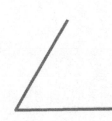

First, draw 2 lines to make a rectangle.

Second, circle all the corners in red.

Third, put an X on the longer sides.

First, draw a line to complete the hexagon.

Second, color the hexagon blue.

Third, write the number of sides the hexagon has in the box below.

On the back of your paper, draw:

- A closed shape with 3 straight sides.
- A closed shape with 4 straight sides.
- A closed shape with 6 straight sides.

Lesson 1:   Describe the systematic construction of flat shapes using ordinal numbers.

© 2018 Great Minds®. eureka-math.org

277

Trace the shapes. Then, use a ruler to draw similar shapes in the large rectangle.

It is easy to trace shapes! I take my time and try to stay on the dashed line!

Hexagons are tricky to draw because they have 6 sides. The sides don't have to be the same length. I know that as long as the shape is closed and has 6 sides, it is a hexagon!

Name _____   Date _____

Trace the shapes. Then, use a ruler to draw similar shapes, on your own, in the large rectangle. Draw more on the back of your paper if you would like!

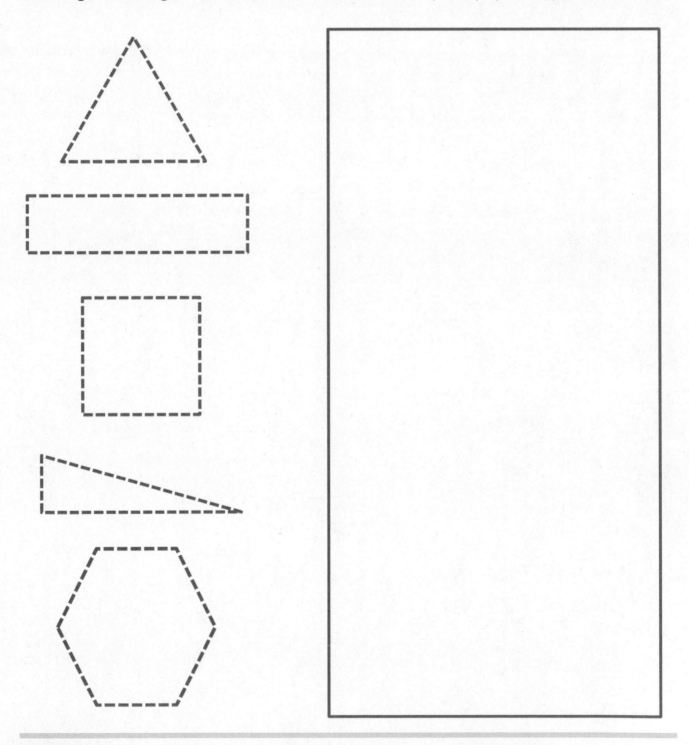

© 2018 Great Minds®. eureka-math.org

Draw something that is a cube.

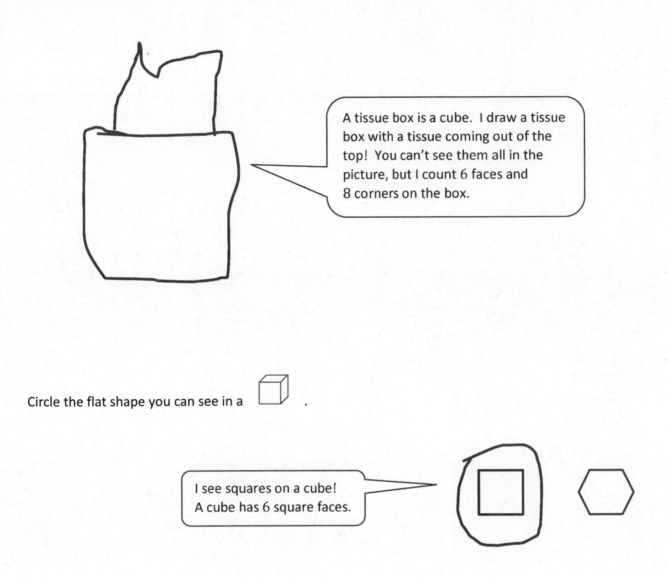

A tissue box is a cube. I draw a tissue box with a tissue coming out of the top! You can't see them all in the picture, but I count 6 faces and 8 corners on the box.

Circle the flat shape you can see in a .

I see squares on a cube! A cube has 6 square faces.

Lesson 3:   Compose solids using flat shapes as a foundation.

283

© 2018 Great Minds®. eureka-math.org

Name _____   Date _____

Draw something that is a cylinder.

Circle the flat shape you can see in a ⬠.

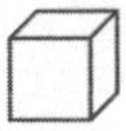

Draw something that is a cube.

Circle the flat shape you can see in a ◻.

EUREKA MATH

© 2018 Great Minds®. eureka-math.org

Draw something that is a cone.

Circle the flat shape you can see in a ▲.

Draw a 3-dimensional solid. Draw one of your solid's faces. Tell an adult about the shapes you drew.

Note to Family Helpers: Your child knows how to name some 3-dimensional solids: cylinders, cones, cubes, and spheres. You can often find these 3-D shapes around the house in objects such as soup cans, ice cream cones, boxes, and balls. For the last question, it is acceptable for your student to find and draw a different type of 3-D solid. Talk about the number of edges, corners, and faces on the object.

Color the 2nd ☆ red.

Color the 4th ☆ blue.

Color the 6th ☆ green.

> The star next to the arrow is the 1st star. That's where I start counting.

➡️ ☆ ★ ☆ ★ ☆ ★ ☆ ☆ ☆ ☆

> I color the 2nd star red. It is easy to find the second star! I just count 2 stars. I do the same thing with the 4th star.

> I can count to 6 to find the 6th star. Or, I can just count on from the blue one, like this: fooouuur, 5, 6.

Name _____   Date _____

Color the 1ˢᵗ ☆ red.
Color the 3ʳᵈ ☆ blue.
Color the 5ᵗʰ ☆ green.
Color the 8ᵗʰ ☆ purple.

Put an X on the 2ⁿᵈ shape.
Draw a triangle in the 4ᵗʰ shape.
Draw a circle around the 6ᵗʰ shape.
Draw a square in the 9ᵗʰ shape.

Draw a circle in the 7ᵗʰ shape.
Put an X on the 1ˢᵗ shape.
Draw a square in the 5ᵗʰ shape.
Draw a triangle in the 3ʳᵈ shape.

Match each animal to the place where it finished the race.

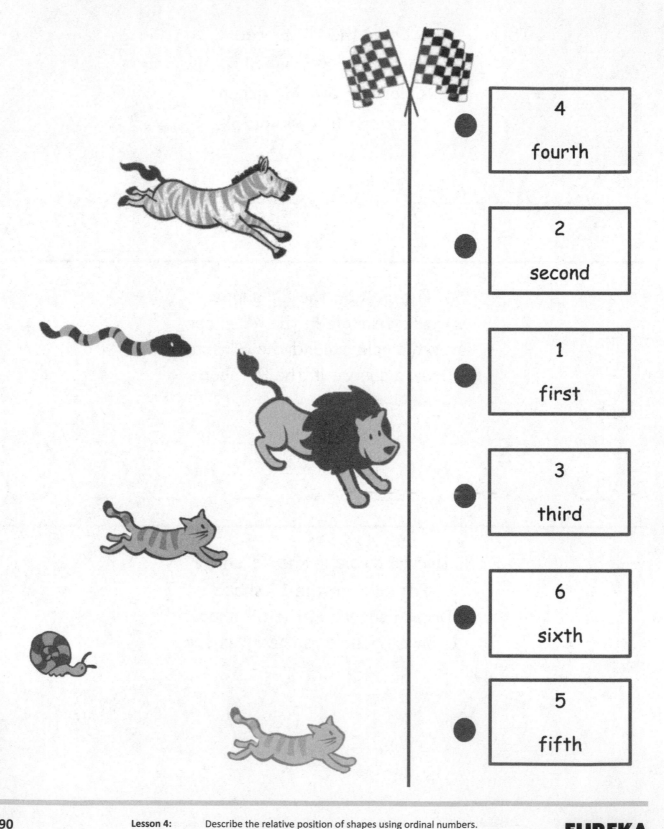

© 2018 Great Minds®. eureka-math.org

Match each group of shapes on the left with the new shape they make when they are put together.

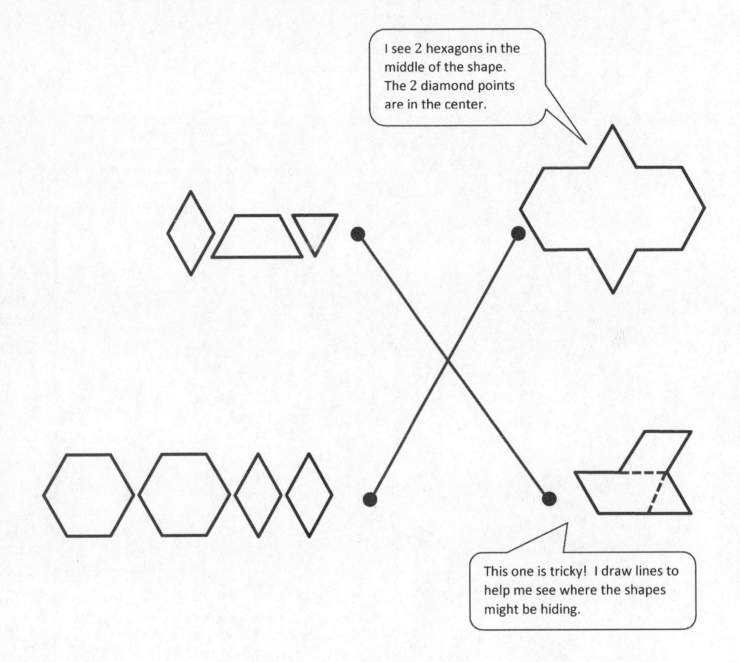

I see 2 hexagons in the middle of the shape. The 2 diamond points are in the center.

This one is tricky! I draw lines to help me see where the shapes might be hiding.

Name _____   Date _____

Match each group of shapes on the left with the new shape they make when they are put together.

Lesson 5:   Compose flat shapes using pattern blocks and drawings.

293

© 2018 Great Minds®. eureka-math.org

Cut out the triangles at the bottom of the paper. Use the small triangles to make the big shape. Draw lines to show where the triangles fit. Count how many small triangles you used to make the big shape.

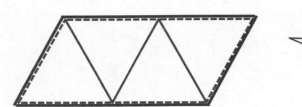

I use 4 of the triangles to make the big shape. I turn them different ways to make them fit. Then, I trace them. It's like the 4 triangles are hiding inside of the big shape!

The big shape is made with ___4___ small triangles.

© 2018 Great Minds®. eureka-math.org

Name _____   Date _____

Cut out the triangles at the bottom of the paper. Use the small triangles to make the big shapes. Draw lines to show where the triangles fit. Count how many small triangles you used to make the big shapes.

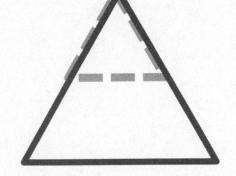

This big triangle is made with _____ small triangles.

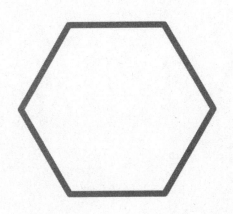

This hexagon is made with _____ small triangles.

- -

© 2018 Great Minds®. eureka-math.org

Using your ruler, draw 2 straight lines from side to side through the shape. Describe to an adult the new shapes you made.

First, I make a straight line across the square. Then, I make another line going from the top of the square to the bottom.

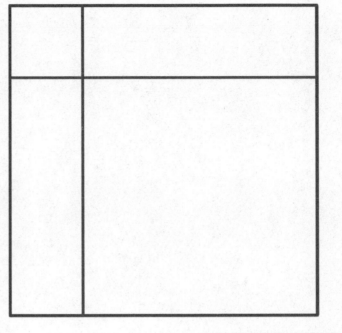

The lines I draw on the square make 4 new rectangles. 2 of the rectangles are squares! It is fun making new shapes!

Lesson 7:   Compose simple shapes to form a larger shape described by an outline.

299

© 2018 Great Minds®. eureka-math.org

Name _____   Date _____

Using your ruler, draw 2 straight lines from side to side through each shape. The first one has been started for you. Describe to an adult the new shapes you made.

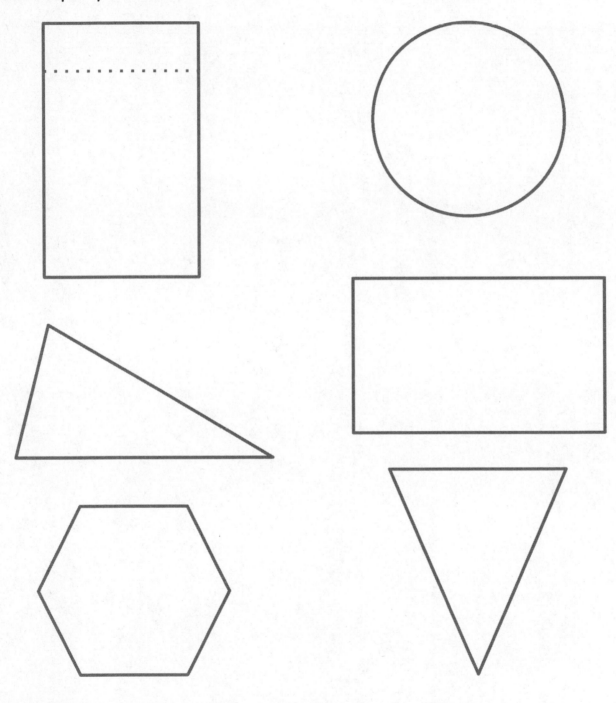

EUREKA
MATH

Lesson 7:   Compose simple shapes to form a larger shape described by an outline.

301

© 2018 Great Minds®. eureka-math.org

Credits

Great Minds® has made every effort to obtain permission for the reprinting of all copyrighted material. If any owner of copyrighted material is not acknowledged herein, please contact Great Minds for proper acknowledgment in all future editions and reprints of this module.